Vic Lewchenko

D1548043

PROTEINS: A THEORETICAL PERSPECTIVE OF DYNAMICS, STRUCTURE, AND THERMODYNAMICS

ADVANCES IN CHEMICAL PHYSICS

VOLUME LXXI

EDITORIAL BOARD

PROTEINS: A THEORETICAL PERSPECTIVE OF DYNAMICS, STRUCTURE, AND THERMODYNAMICS

CHARLES L. BROOKS III

Department of Chemistry,
Carnegie-Mellon University,
Pittsburgh, Pennsylvania

MARTIN KARPLUS

Department of Chemistry,
Harvard University,
Cambridge, Massachusetts

B. MONTGOMERY PETTITT

Department of Chemistry
University of Houston
Houston, Texas

ADVANCES IN CHEMICAL PHYSICS
VOLUME LXXI

Series editors

Ilya Prigogine

*University of Brussels
Brussels, Belgium
and
University of Texas
Austin, Texas*

Stuart A. Rice

*Department of Chemistry
and
The James Franck Institute
University of Chicago
Chicago, Illinois*

WILEY

AN INTERSCIENCE® PUBLICATION
JOHN WILEY & SONS
NEW YORK • CHICHESTER • BRISBANE • TORONTO • SINGAPORE

An Interscience® Publication

Library of Congress Cataloging-in-Publication Data:
Brooks, Charles.
  Proteins: a theoretical perspective of dynamics,
structure, and thermodynamics

  (Advances in chemical physics; v. 71)
  "An Interscience publication."
  Bibliography: p.
  Includes index.
  1. Proteins—Structure.   2. Thermodynamics.
3. Biophysics.   I. Karplus, Martin, 1930-
II. Pettitt, B. Montgomery.   III. Title.   IV. Series.

QD453.A27   vol. 71 [QP551]   539 s [547.7'5]   87-15993
ISBN 0-471-62801-8

Printed in the United States of America

10 9 8 7 6 5 4 3 2 1

To Aneesur Rahman, a pioneer in molecular dynamics simulations; his scientific and personal contributions are an inspiration to all who work in this area.

INTRODUCTION

Few of us can any longer keep up with the flood of scientific literature, even in specialized subfields. Any attempt to do more and be broadly educated with respect to a large domain of science has the appearance of tilting at windmills. Yet the synthesis of ideas drawn from different subjects into new, powerful, general concepts is as valuable as ever, and the desire to remain educated persists in all scientists. This series, *Advances in Chemical Physics,* is devoted to helping the reader obtain general information about a wide variety of topics in chemical physics, a field which we interpret very broadly. Our intent is to have experts present comprehensive analyses of subjects of interest and to encourage the expression of individual points of view. We hope that this approach to the presentation of an overview of a subject will both stimulate new research and serve as a personalized learning text for beginners in a field.

ILYA PRIGOGINE
STUART A. RICE

PREFACE

The long-range goal of molecular approaches to biology is to describe living systems in terms of chemistry and physics. Dirac's often quoted statement (*Proc. Roy. Soc.* [London] **123**, 714 [1929])

> The underlying physical laws necessary for the mathematical theory of a large part of physics and the whole of chemistry are thus completely known, and the difficulty is only that the exact application of these laws leads to equations much too complicated to be soluble.

is equally applicable to biology. Great progress has been made over the last thirty years in applying the equations to chemical problems involving the structures and reactions of small molecules. It is only recently, however, that corresponding studies have been undertaken for the mesoscopic systems of importance to biology. One essential step is to express the properties of macromolecules, such as proteins, in terms of the component atoms and the force laws governing their interactions. For one who had spent a part of his career interpreting elementary chemical reactions, it was an exciting prospect to extend the dynamical methods that had been successfully applied to simple reactions to the internal motions of systems composed of thousands of atoms. One of the original objectives of such calculations, not yet fully realized, was to describe enzyme catalyzed reactions at the same level of detail as the well-studied H + H$_2$ exchange reaction.

This volume shows how molecular dynamics simulations can be used to provide information concerning the structure, dynamics, and thermodynamics of biologically interesting macromolecules. The potential surfaces on which the atomic motions take place are ultimately determined by the Schrödinger equation, although empirical approximations are used in most cases to make the calculations tractable. As to the motions themselves, they can be treated classically at room temperature by solving Newton's equations. It is fitting, therefore, that this review should be written in 1987—the 300th anniversary of the publication of Newton's *Principica Mathematica* and the 100th anniversary of Schrödinger's birth. Schrödinger would have been pleased to see some of the detailed biological applications of the basic laws of physics that are presaged in his influential book "What Is Life."

When the first studies of the molecular dynamics of protein were made a little over ten years ago, both chemists and biologists expressed their feeling that the calculations were a waste of time—chemists felt that detailed treatments of such complex systems were impossible, and biologists believed, that even if they were possible, it would add little, if anything, of importance to our knowledge. Experience has proved the contrary. There has been a very rapid development in molecular dynamics simulations that are providing a basis for a more complete understanding of these macromolecules and are aiding in the interpretation of experiments concerned with their properties. Although such studies do not, in themselves, constitute a theoretical approach to biology, they present an important contribution to our detailed knowledge of the essential components of living systems. In fact, so many publications have already appeared in the area of macromolecular simulations that it is difficult to include everything that has been done in a single volume. We have attempted to include most of the important examples from the theoretical literature, although we do emphasize our own results. Experimental studies are introduced as they relate to specific aspects of the theory.

This volume is a review for the series *Advances in Chemical Physics*. As such, it draws heavily on earlier reviews by the authors and does not pretend to be a textbook in the field of protein dynamics. Nevertheless, it will hopefully serve a useful function by introducing both chemists and physicists to this exciting field, in addition to providing a relatively up-to-date review for those working in this rapidly developing area.

We would like to thank our many collaborators, who spent some time at Harvard, for helping to make the field of macromolecular dynamics what it is today. Some of them have also contributed by reading various portions of this volume. They are: P. Bash, D. Bashford, A. Beyer, J. Brady, B. R. Brooks, R. E. Bruccoleri, A. T. Brünger, R. Carlson, D. A. Case, L. Caves, R. D. Coalson, F. Colonna, S. Cusack, P. Derreumaux, C. M. Dobson, R. Elber, M. Field, J. Gao, B. Gelin, D. A. Giammona, W. F. van Gunsteren, H. Guo, K. Haydock, J. C. Hoch, B. Honig, V. Hruby, R. Hubbard, T. Ichiye, K. K. Irikura, K. Kuczera, J. Kuriyan, J. N. Kushick, A. W.-M. Lee, S. Lee, R. M. Levy, C. Lim, J. A. McCammon, A. Miranker, S. Nakagawa, D. Nguyen, J. Novotny, B. D. Olafson, E. T. Olejniczak, D. Perahia, C. B. Post, W. Reiher, C. M. Roberts, P. J. Rossky, B. Roux, N. Sauter, H. H.-L. Shih, J. Smith, D. J. States, J. Straub, N. Summers, S. Swaminathan, A Szabo, B. Tidor, W. Van Wesenbeeck, A. Warshel, D. Weaver, M. A. Weiss, P. G. Wolynes, and H.-A. Yu.

Thanks also go to Cheri Brooks, Patricia E. Gleason, and Marci Karplus for help with typing, checking references, and correcting errors.

Martin Karplus

Manigod, France
December, 1987

CONTENTS

PROTEINS: A THEORETICAL PERSPECTIVE OF DYNAMICS, STRUCTURE, AND THERMODYNAMICS

ADVANCES IN CHEMICAL PHYSICS

VOLUME LXXI

CHAPTER I

INTRODUCTION

Proteins are one of the essential components of living systems. Along with nucleic acids, polysaccharides, and lipids, proteins constitute the macromolecules that have important roles in biology. Nucleic acids, in the form of DNA and RNA, store and distribute the genetic information as needed. Of particular importance is the information that determines the sequences of amino acids that characterize the proteins. Proteins contribute to the structure of an organism and execute most of the tasks required for it to function. Proteins even form part of the complex mechanism by which they are synthesized. Polysaccharides, linear and branched-chain polymers of sugars, provide structural elements, store energy, and when combined with peptides or proteins, play an important role in antigenicity and, more generally, in cellular recognition. Lipids, which include molecules such as fatty acids, phospholipids, and cholesterol, serve as energy sources and are the most important components of the membrane structures that organize and compartmentalize cellular function.

In this volume we concentrate on globular proteins, the biological macromolecules with the greatest functional range. It is for these systems that the relation of function to structure and dynamics is best understood. Most chemical transformations that occur in living systems are catalyzed by enzymes, the globular proteins that have evolved for executing such specific tasks. As well as enhancing the rates of reactions, sometimes by eight or more orders of magnitude, globular proteins (e.g., repressors) inhibit certain reactions (e.g., the transcription of DNA) involved in the mechanism for the control of growth and differentiation. A breakdown of these control mechanisms can lead to unobstructed growth and the development of cancer. Other proteins (such as hemoglobin) serve to transport small molecules (such as oxygen), electrons, and energy to the appropriate parts of the organism. Antibody molecules are proteins that protect the organism by specifically recognizing and binding to foreign antigenic substances (such as viruses). Many proteins have structural roles; e.g., fibrous tissue is composed mainly of the protein collagen, and the major functional components of muscle, actin and myosin, are proteins.

Because of this wide range of protein functions and the need to develop

1

specialized proteins for each of them, the number of different proteins in an organism can be very large. The well-studied bacterium *Escherichia coli* contains about 3000 different kinds of proteins. Since many of them occur in multiple copies, there are a total of about 1 million protein molecules in a single bacterium. In human beings there are estimated to be on the order of 10^5 to 10^6 different proteins.

For most globular proteins, the biological function includes an interaction with one or more small molecules (a ligand, hormone, substrate, coenzyme, chromophore, etc.) or another macromolecule. Whether reactive or nonreactive systems are being considered, there can be important conformational alterations in the molecule that is bound and concomitant changes in the structure of the macromolecule to which the binding occurs. Such concerted conformational changes are the essential element for activity in some cases; in others, they play a less significant role. In hormone-receptor binding, for example, the structural changes induced in the receptors are fundamental to the transmission of information. Correspondingly, the conformational transition induced by ligand binding in hemoglobin is an integral part of the cooperative mechanism. Further, in many systems, small motions have been observed (e.g., the differences between the ligated and unligated structure of ribonuclease A) that appear to be involved in the function of the protein. Thus any attempt to understand the details of the activity of proteins requires an investigation of the dynamics of the structural fluctuations and their relation to reactivity and conformational change.

In addition to their biological importance, globular proteins are intrinsically interesting systems from the viewpoint of physical chemistry. They are long-chain polymers, but unlike most polymers they have a well-defined average structure. This structure is aperiodic (the "aperiodic crystal" of Schrödinger)[1] in the sense that it does not have regular repeats. Since the structure is determined by weak, noncovalent, interactions among the elements of the polypeptide chain, large fluctuations are expected. For a complete description of proteins, it is important, therefore, to know, in addition to the average structure, the form of the fluctuations that occur, to determine how they take place, and to evaluate their magnitudes and time scales.

Historically, hydrogen exchange experiments (i.e., the replacement of one isotope of hydrogen bound to an O, N, or S atom in the protein interior by another isotope from the solvent water) provided some of the earliest evidence for the existence of conformational fluctuations in proteins. More recently, a wide range of experimental methods (such as fluorescence quenching and depolarization, nuclear magnetic resonance relaxation, infrared and Raman spectroscopy, and X-ray and inelastic neutron scattering) have been used to study the motions in proteins. However, it is primarily the application of theoretical methods, particularly molecular dynamics simulations, that have brought about a conceptual change in the pervading view concerning the nature of proteins.

Although to chemists and physicists it is self-evident that polymers such as proteins undergo significant fluctuations at room temperature, the classic view of such molecules in their native state had been static in character. This followed from the dominant role of high-resolution X-ray crystallography in providing structural information for these complex systems. The remarkable detail evident in crystal structures led to an image of biomolecules with every atom fixed in place. Tanford suggested that as a result of packing considerations "the structure of proteins must be quite rigid."[2] D. C. Phillips, who determined the first enzyme crystal structure, has written: "The period 1965–75 may be described as the decade of the rigid macromolecule. Brass models of DNA and a variety of proteins dominated the scene and much of the thinking."[3] Molecular dynamics simulations have been instrumental in changing the static view of the structure of biomolecules to a dynamic picture. It is now recognized that the atoms of which biopolymers are composed are in a state of constant motion at ordinary temperatures. The X-ray structure of a protein provides the average atomic positions, but the atoms exhibit fluidlike motions of sizable amplitudes about these averages. Crystallographers have acceded to this viewpoint and have come so far as sometimes to emphasize the parts of a molecule they do not see in a crystal structure as evidence of motion or disorder. The new understanding of protein dynamics subsumes the static picture. Knowledge of the average atomic positions allows discussion of many aspects of biomolecule function in the language of structural chemistry. However, recognition of the importance of fluctuations opens the way for more sophisticated and accurate interpretations of protein activity.

Simulations of proteins, as of many other systems (e.g., liquids), can, in principle, provide the ultimate details of motional phenomena. The primary limitation of simulation methods is that they are approximate. It is here that experiment plays an essential role in validating the simulation methods; that is, comparisons with experimental data serve to test the accuracy of the calculated results and provide criteria for improving the methodology. However, the experimental approaches to biomolecular dynamics are limited as to the information that can be obtained from them; e.g., if one is concerned with the time scale of motions, the frequency spectrum covered by experiments such as nuclear magnetic resonance (NMR) is incomplete, so that motional models that are able to rationalize the data can be inaccurate. When experimental comparisons indicate that the simulations are meaningful, their capacity for providing detailed results often makes it possible to examine specific aspects of the atomic motions far more easily than by making measurements. However, at the present stage of development, possible inaccuracies in the simulations must be kept in mind in evaluating and applying the results.

The present volume deals primarily with theoretical approaches to protein dynamics and thermodynamics. This rapidly developing field of study is founded on efforts to supplement our understanding of protein structure with concepts and techniques from modern chemical theory, including reaction dynamics and quantum and statistical mechanics. From a knowledge of the potential energy surface for a protein, the forces on the component atoms can be calculated and used to determine the phase space trajectory for the molecule at a given temperature. Such molecular dynamics simulations, which have successfully been applied to gases and liquids containing a large number of atoms, provide information concerning the thermodynamic properties and the time dependence of processes in the system of interest. More generally, statistical mechanical techniques are being used widely to characterize molecular motions and chemical reactions in condensed phases. The application of these methods to protein molecules is natural in that proteins contain many atoms, are densely packed, and function typically in a liquid environment.

Before focusing on the dynamical studies of biomolecules, it is useful to place this new field in perspective relative to the more general development of molecular dynamics. Molecular dynamics has followed two pathways which come together in the study of biomolecule dynamics. One of these, usually referred to as trajectory calculations, has an ancient history that goes back to two-body scattering problems for which analytic solutions can be achieved. However, even for only three particles with realistic interactions, difficulties arise. An example is provided by the simplest chemical reaction, $H + H_2 \rightarrow H_2 + H$, for which a prototype calculation was attempted by Hirschfelder, Eyring, and Topley in 1936.[4] They were able to calculate a few steps along one trajectory. It was nearly 30 years later that the availability of computers made it possible to complete the calculation.[5] Much has been done since then in applying classical trajectory methods to a wide range of chemical reactions.[5-7] These classical studies have been supplemented by semiclassical and quantum-mechanical calculations in areas where quantum effects can play an important role.[7,8] The focus at present is on more complex molecules, the redistribution of their internal energy, and the effect of this on reactivity.[9]

The other pathway in molecular dynamics has been concerned with physical rather than chemical interactions (in analogy to physisorption versus chemisorption) and with the thermodynamic and dynamic properties of a large number of particles, rather than detailed trajectories of a few particles. Although the basic ideas go back to van der Waals and Boltzmann, the modern era began with the work of Alder and Wainright on hard-sphere liquids in the late 1950s.[10] The paper by Rahman[11] in 1964, on a molecular dynamics simulation of liquid argon with a soft sphere (Lennard-Jones) potential represented an important next step. Simulations of complex liquids followed; the now classic study of liquid water by Stillinger and Rahman was published in

1974.[12] Since then there have been many studies on the equilibrium and non-equilibrium behavior of a wide range of systems.[13,14]

This background set the stage for the development of molecular dynamics of biomolecules. The size of an individual molecule, composed of 500 or more atoms for even a small protein, is such that its simulation in isolation can serve to obtain approximate equilibrium properties, as in the molecular dynamics of fluids, although detailed aspects of the atomic motions are of considerable interest, as in trajectory calculations. A basic assumption in initiating such studies was that potential functions could be constructed which were sufficiently accurate to give meaningful results for systems as complex as proteins or nucleic acids. In addition, it was necessary to assume that for these inhomogeneous systems, in contrast to the homogeneous character of even "complex" liquids such as water, simulations of an attainable time scale (10 ps in the initial studies) could provide a useful sample of the phase space in the neighborhood of the native structure. For neither of these assumptions was there strong supporting evidence in 1975. Nevertheless, the techniques of molecular dynamics were employed with the available potential functions in the first simulation of the internal atomic motions of a protein, the bovine pancreatic trypsin inhibitor (BPTI),[15] which has played the role of the "hydrogen molecule" of protein dynamics.

In this volume we summarize first the elements of protein structure and provide a brief overview of the internal motions of proteins, their relation to the structural elements, and their functional role. We then outline the theoretical methods that are being used to study motional phenomena and thermodynamics. A description is given of the potential functions that determine the important interactions, and the various approaches that can be used to study the dynamics are outlined. Since the motions of interest involve times from femtoseconds to seconds or longer, a range of dynamical methods is required.

An important consideration in protein dynamics is the influence of solvents such as water on the functional integrity and structural stability of the biomolecular system. This influence is manifested in a variety of different phenomena, ranging from marked solvent effects on the rate of oxygen uptake in myoglobin to the stabilization of oppositely charged sidechain pairs on the surface of proteins. Although experimental data on protein-solvent interactions are being accumulated, our understanding of the structural, dynamic, and thermodynamic effects of water on biological systems is still incomplete. Some of the newer developments in the theory of aqueous solutions are described and it is shown how they can help to provide a fundamental understanding of solvated proteins.

Studies of the dynamics are of utility for determining thermodynamic properties as well as for providing information concerning the motions themselves. Of special interest is an understanding of the stability of proteins and the thermodynamics of their interactions with drugs and ligands. Theoretical methods are described for determining the free energies involved. Since the phenomena occur in the liquid state or some other condensed phase, it is necessary to be able to include the effect of solvent in going from the microscopic interactions to the macroscopic enthalpies, entropies, and free energies that are the experimental thermodynamic variables of interest. Such information, when augmented by the results of special techniques for the study of chemical reactions, leads naturally to an analysis of the reaction dynamics involving macromolecules.

The main body of this volume presents results that have been obtained in dynamical studies of proteins in vacuum, in solution, and in crystals. Because of the intense activity in this area, a selection has been made to provide a representative and coherent view of our present knowledge. Where possible, comparisons with experiment and the functional correlates of the motions are stressed. A description is given of specific experimental areas that are of particular importance for the analysis of dynamics or where the simulation results are providing information essential for the interpretation of the experimental data. We conclude with an outlook for future developments and applications in this exciting field.

A number of reviews on related material in protein dynamics have appeared. For reviews concerned primarily with theoretical work, the reader may read Careri et al.,[16,17] Cooper,[18,19] Weber,[20] Karplus,[21] Karplus and McCammon,[22] Levitt,[23] Levy,[23a] McCammon and Karplus,[24,25] McCammon,[26] Pettitt and Karplus,[26a] van Gunsteren and Berendsen,[27] and Welch et al.[28] Experimental work is reviewed in Campbell et al.,[29] Cusack,[29a] Debrunner and Frauenfelder,[30] Dobson and Karplus,[30a] Englander and Kallenbach,[31] Gurd and Rothgeb,[32] Karplus and McCammon,[33] Jardetsky,[34] Bennett and Huber,[35] Peticolas,[36] Ringe and Petsko,[37] Torchia,[37a] Williams,[38,39] Wagner and Wüthrich,[39a] and Woodward and Hilton.[40] In addition, several volumes reporting on meetings devoted to protein dynamics have been published,[41,42] as has an article in *Scientific American*.[43] An introductory description of the dynamics of proteins and nucleic acid has been presented by McCammon and Harvey.[44]

CHAPTER II

PROTEIN STRUCTURE AND DYNAMICS— AN OVERVIEW

Since the function and dynamics of proteins are intimately related to their structure, we first provide a short survey of the structural elements of proteins. We then outline briefly the present state of our knowledge of protein dynamics and the role of the internal motions in protein function.

A. THE STRUCTURE OF PROTEINS

Much of what we now know about the structure of globular proteins comes from X-ray crystallographic studies.[45-48] The first high-resolution structures of proteins were those of myoglobin and hemoglobin determined in the early 1960s by J. Kendrew and coworkers[49] and M. F. Perutz,[50] respectively, and the first structure of an enzyme, lysozyme, was reported by D. C. Phillips and coworkers in 1965.[51] Since then a large number of protein structures have been determined. A recent listing (1987) of the Protein Data Bank at Brookhaven National Laboratories (where the results of many but not all protein structure determinations are deposited)[52] includes the coordinates for about 200 different proteins. These results have made possible an analysis of the "anatomy" of protein structures.[53] We show in Fig. 1 a schematic view of the structures of several proteins that illustrate some of the structural motifs that have been found.

Each protein consists of a polypeptide chain that is made up of residues or amino acids linked together by peptide bonds. The polypeptide chain backbone, a portion of which is shown in Fig. 2, is composed of repeating units that are identical, except for the chain termini. Proteins vary widely in size, from 50 to 500 or so residues, corresponding to 1000 to 10,000 or so atoms. Approximately half of the atoms are hydrogens, which are not seen except in very high resolution X-ray and in neutron crystal structures; thus, most of the descriptions of proteins focus on the positions of the "heavy" atoms, C, N, O, and S. What distinguishes different proteins, other than the number of amino acids, is the sequence of amino acids in the polypeptide chain. There are 20

7

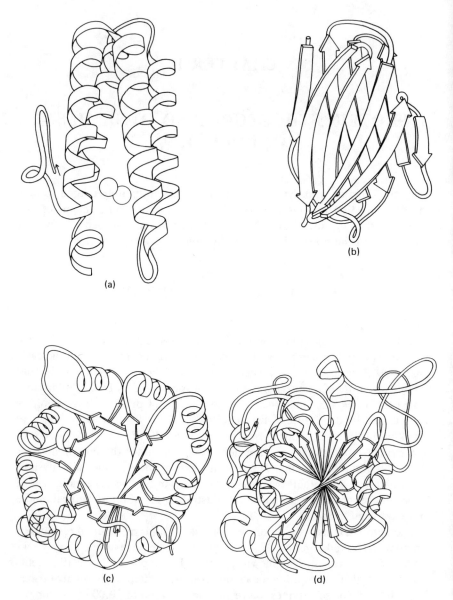

Figure 1. Schematic representations of protein structures: (*a*) myohemerythrin, an α-helical protein with antiparallel helices; (*b*) V₂ domain of an immunoglobulin, a β-sheet protein; (*c*) triose phosphate isomerase, a parallel α-β protein with a central "β barrel"; (*d*) carboxypeptidase, a parallel α-β protein with a central β-sheet structure; (*e*) *para*-hydroxybenzoate hydrolase, a complex protein structure with more than one domain. (From Ref. 53; courtesy of J. Richardson.)

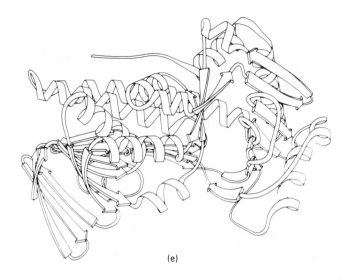

(e)

commonly occurring amino acids that differ in their sidechains; they vary from the simplest, glycine, to the most complex, tryptophan (Fig. 3).

It is the sequence of amino acids, referred to as the primary structure of the protein, that determines the native conformation, the structure that is stable under physiological conditions. The first protein amino acid sequence was determined by F. Sanger and coworker in 1953 for insulin.[54] It is generally believed that the native structure corresponds to a free-energy minimum, although there is no direct experimental or theoretical evidence for this. One suggestive result is that it is possible to denature (unfold) many proteins in

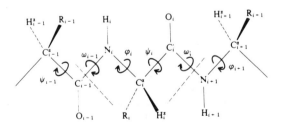

Figure 2. Polypeptide backbone with standard notation for mainchain atoms; the flexible dihedral angles ϕ and ψ and the more rigid, partially conjugated peptide bond angle ω are shown. The sidechains are indicated as R. [Adapted from C. R. Cantor and P. R. Schimmel, *Biophysical Chemistry* (W. H. Freeman and Co., San Francisco, 1980).]

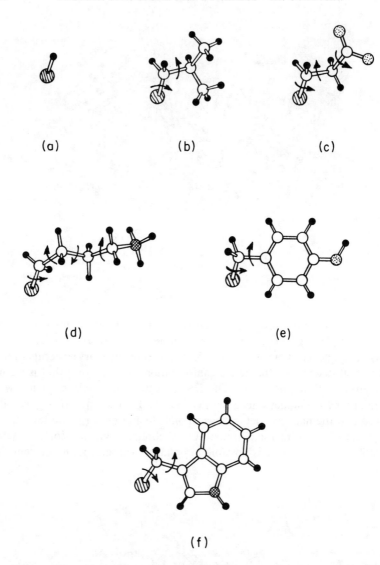

Figure 3. Some of the commonly occurring sidechains: (*a*) glycine (Gly, G); (*b*) leucine (Leu, L); (*c*) glutamic acid (Glu, E); (*d*) lysine (Lys, K); (*e*) tyrosine (Tyr, Y); (*f*) tryptophan (Trp, W). All atoms, including hydrogens (black dots), are shown; the α-carbons are indicated by lines, oxygens by dots, and nitrogens by crosshatching in the corresponding circles; open circles are other carbon atoms; and the flexible dihedral angles are indicated by arrows with the one nearest to the α-carbon called χ^1, the next χ^2, and so on.

solution by increasing the temperature or lowering the pH and then to recover the native protein by returning the solution to normal values of temperature or pH.[55]

From the analysis of many protein structures, it is found that the folding of portions of the polypeptide chain often has certain regularities, called elements of secondary structure. These can be defined in terms of the main-chain dihedral angles ϕ and ψ (see Fig. 2); it is not necessary to delimit the angle ω associated with the partially conjugated peptide bond because it is generally in the neighborhood of $180 \pm 5°$. The most important secondary structural elements are the α-helix (Fig. 4a) and β-pleated sheets (Fig. 4b), both of which are regularly repeating structures with backbone hydrogen bonds that were predicted by Pauling and Corey[56] in the 1950s, prior to the first protein structure determination. The α-helix is a compact rodlike structure with 3.6 amino acids per turn, a rise of only 1.5 Å per turn, and a $C=O\cdots H-N$ hydrogen bond between residues i and $(i + 4)$. Wool (α-keratin) has an α-helix as its essential constituent and because it is so compact leads to the well-known extensibility of that fiber. Other helical structures [e.g., the 3_{10} helix with 3 residues per turn and an i-to-$(i + 3)$ hydrogen bond] also occur in proteins. The β-pleated sheet structure is an extended structure with a displacement of approximately 3.47 Å per residue (Fig. 4b). It can be regarded as a (degenerate) twofold helix in which the hydrogen bonds are between strands rather than within a strand, as for the α-helix. Pleated sheets can be formed with parallel or antiparallel orientations of adjacent strands. Silk, which is formed from antiparallel β-pleated sheets, is a very strong but rigid fiber because the strands are already extended to near their maximum length. Since globular proteins have a finite size with radii of gyration from 15 to 80 Å or so, the secondary structural elements, such as α-helices and β-sheets, are limited in length. They often terminate in so-called turns, which have also been shown to have regular features that can be classified into a number of types (Fig. 4c).

When a large number of protein structures are examined, it is found that on the average, 25% of the amino acids are in helices, 25% in sheets, 25% in turns, and the remaining 25% in what are called random coil segments that have no simple regularity in their mainchain dihedral angles. A given protein structure may deviate widely from this set of averages, however; e.g., myoglobin has 85% of the residues in the α-helical configuration and the remainder in turns or random-coil sequences.

The overall spatial arrangements of the amino acid residues in proteins are referred to as the tertiary structure. In many cases this can be described approximately in terms of the packing together of secondary structural elements. Various motifs have been identified; these include the helix-turn-helix structure and the β strand–α helix–β strand structure, as well as others (see

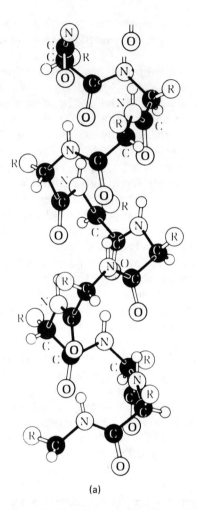

(a)

Figure 4. Protein secondary structural elements: (a) right-handed α-helix showing intra-chain hydrogen bonds as dotted lines (α_R; $\phi \simeq -60°$, $\psi \simeq -60°$); (b) antiparallel β-pleated sheet showing interchain hydrogen bonds as dashed lines (β_A: $\psi \simeq -120°$, $\psi \simeq 120°$); (c) β-turns of types I and II, differing in the orientation of the central peptide group. [Part (a) is adapted from A. L. Lehninger, *Biochemistry* (Worth Publishers, Inc., New York, 1975); (b) from Ref. 81; and (c) from Ref. 53.]

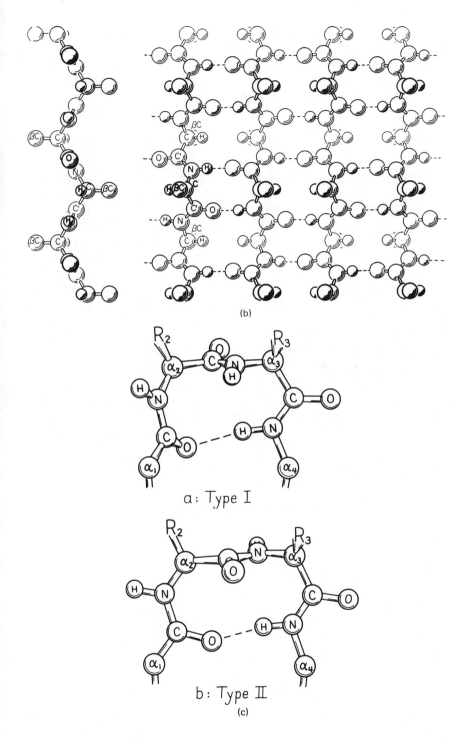

(b)

a: Type I

b: Type II

(c)

13

Fig. 1). However, there also occur regions in proteins involving random-coil segments (e.g., some proteins have essentially no identifiable secondary structure) that require a more complex description. In such cases it may not be possible to do more than simply give a list of the mainchain dihedral angles (ϕ_i, ψ_i for all residues i).

Even in the absence of secondary structure, there are regularities in the amino acid distributions and their packing. In general, proteins are tightly packed systems (approximately the packing density of close-packed spherical atoms) with only a few voids (Fig. 5). Nonpolar sidechains of amino acids tend to be in the interior of the protein, often with clusters of aromatic and other nonpolar residues forming a stabilizing core. Most charged sidechains are on the surface of the protein, with clusters of such amino acids often concentrated in the active site. Polar residues, as well as the carbonyl and amide groups of the polypeptide chain, tend to be more uniformly distributed, with essentially all hydrogen-bond donors or acceptors located so that they form hydrogen bonds either with other parts of the protein or with the surrounding solvent.

In some cases proteins are divided into two or more domains (Fig. 1), each of which is like a globular protein but connected covalently to other domain(s) by the continuous polypeptide chain. Other proteins are oligomeric in that they are composed of several unconnected polypeptide chains (subunits) that usually, but not always, fold up independently and assemble to form the complete protein. The arrangement of the subunits relative to each other is referred to as the quaternary structure. Hemoglobin ($\alpha_2\beta_2$) (Fig. 6) and aspartate transcarbamoylase ($\alpha_6\beta_6$), where α and β refer to different types of subunits, are well-studied cases where different quaternary structures occur with significantly altered properties.

B. OVERVIEW OF PROTEIN MOTIONS

The general motional characteristics of globular proteins follow directly from their structural properties. The polypeptide chain of the protein has strong covalent bonding forces along the chain but relatively weak, noncovalent interactions between different parts of the chain that are packed together in the native structure; a few of the noncovalent interactions involve charged groups that form "salt links" whose interaction energy can approach that of a covalent bond. The only covalent interactions between different parts of the chain in globular proteins are disulfide bonds formed by oxidation of pairs of cystine sidechains, of which there are usually no more than 1 for every 20 or so residues. Fibrous proteins, such as elastin, have cross-links involving sidechains such as lysine.

The polypeptide chain of a protein has single bonds that permit internal torsional rotation to take place. This is true for the ϕ and ψ angles of each

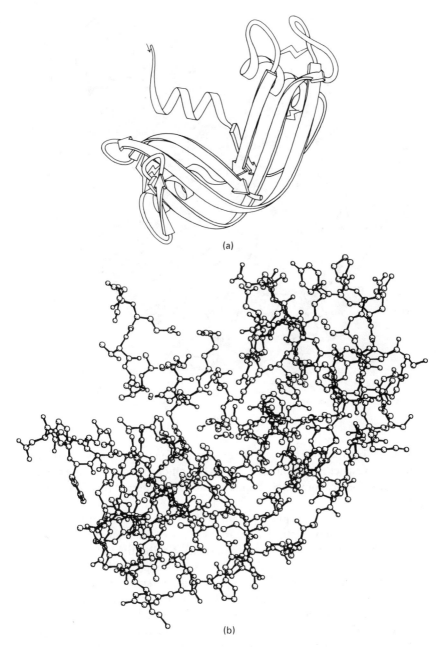

(a)

(b)

Figure 5. Ribonuclease A: (*a*) schematic diagram; (*b*) diagram showing heavy atoms with small radii; (*c*) diagram showing all heavy atoms with van der Waals radii. [Part (a) was prepared by J. Richardson; (b) and (c) were prepared by A. Brünger from coordinates supplied by G. Petsko.]

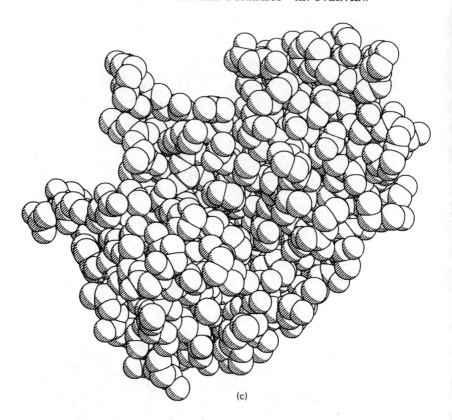

(c)

Figure 5. (*Continued*)

amino acid, with only the peptide groups torsional angle ω being relatively rigid with respect to twisting (Fig. 2). Also, all of the sidechains, except glycine, have one or more single bonds about which internal rotation can occur (Fig. 3).

At any given time, a typical protein exhibits a wide variety of motions; they range from irregular elastic deformations of the entire protein driven by collisions with solvent molecules to seemingly chaotic librations of interior groups driven by collisions with neighboring atoms in the protein. Considering only typical motions at physiological temperatures, the smallest effective dynamical units in proteins are those that behave nearly as rigid bodies because of their covalent bonding. Examples include the phenyl group in the sidechain of tyrosine (Fig. 3), the isopropyl group in the sidechains of valine or leucine (Fig. 3), and the amide groups of the protein backbone (Fig. 2). Except for the methyl rotations in the isopropyl group, these units display only relatively small internal motion, owing to the high energy cost associated with deforma-

Figure 6. Diagram of the quaternary structure of the hemoglobin tetramer showing the four primarily helical subunits and the heme group in each. (Adapted from Ref. 263.)

tions of bond lengths, bond angles, or dihedral angles about multiple bonds. The important motions in proteins involve relative displacements of such groups associated with torsional oscillations about the rotationally permissive single bonds that link the groups together. High-frequency vibrations do occur within the local groups, but these are not of primary importance in the relative displacements.

Most groups in a protein are tightly encaged by atoms of the protein or of the surrounding solvent. At very short times ($\leq 10^{-12}$ s), the groups may display rattling motions in their cages, but such motions are of relatively small amplitude (≤ 0.2 Å). More substantial displacements of the groups occur over longer time intervals; these motions involve concomitant displacements of the cage atoms. Broadly speaking, such "collective" motions may have either a local or a rigid-body character. The former involves changes of the cage structure and relative displacements of neighboring groups, while the latter involves relative displacements of different regions of the protein but only small changes on a local scale.

The presence of such motional freedom implies that a native protein at room temperature samples a range of conformations. Most are in the general neighborhood of the average structure, but at any given moment an individual protein molecule is likely to differ significantly from the average structure. This in no way implies that the X-ray structure, which corresponds to the average in the crystal, is not important. Rather, it suggests that fluctuations about the average can play a significant role in protein function. In a protein, as in any polymeric system in which rigidity is not supplied by covalent cross-links, relatively large-scale fluctuations cannot be avoided. Thus, it is possible that they have a functional role as a consequence of evolutionary development.

Although the existence of fluctuations is now well established, our understanding of their biological role in most areas is incomplete. Both conformational and energy fluctuations with local to global character are expected to be important. In a protein, as in other nonrigid condensed systems, structural changes arise from correlated fluctuations. Perturbations, such as ligand binding, that produce tertiary or quaternary alterations, do so by introducing forces that bias the fluctuations in such a way that the protein makes a transition from one structure to another. Alternatively, the fluctuations can be regarded as searching out the path or paths along which transitions take place.

In considering the internal motions of proteins, one must separate the dynamic from the thermodynamic elements; in the latter, the presence of flexibility is important (e.g., entropy of binding), while in the former the directionality and time scales play a role. Another way of categorizing the two aspects is that in thermodynamics, the equilibrium behavior is the sole concern, while in dynamics, the displacements from the average structure are the essential element. In certain cases, some features of the dynamics may be unimportant because they proceed on a time scale that is faster than the phenomenon of interest. An example might be the fast local relaxation of atoms involved in a much slower hinge-bending motion experienced in an enzyme active site; here only the time scale of the latter would be expected to be involved in determining the important rate process (e.g., product release), although the nature of the former would be of considerable interest. In other situations, the detailed aspects of the atomic fluctuations are a significant factor. This may be the case in the oxygen transport protein myoglobin, where local sidechain motions appear to be essential for the entrance and exit of ligands.

To summarize the available results concerning the dynamics of proteins and their functional role, we present in Table I some examples of the different categories of internal motions that have been identified. They cover a wide range of amplitudes (0.01 to 100 Å), energies (0.1 to 100 kcal/mol) and time scales (10^{-15} to 10^3 s). One expects an increase in one quantity (e.g., the amplitude of the fluctuations) to correspond to an increase in the others (e.g., a larger energy and longer time scale). This is often true, but not always. Some motions are slow because they are intrinsically complex, involving the correlated displacements of many atoms. An example might be partial-to-total un-

TABLE I
Internal Motions of Globular Proteins

I. *Local Motions* (0.01 to 5 Å, 10^{-15} to 10^{-1} s)

   (a) Atomic Fluctuations

      1. Small displacements required for substrate binding (many enzymes)
      2. Flexibility required for "rigid-body" motion (lysozyme, liver alcohol dehydrogenase, l-arabinose binding protein)
      3. Energy "source" for barrier crossing and other activated processes
      4. Entropy "source" for ligand binding and structural changes

   (b) Sidechain Motions

      1. Opening pathways for ligand to enter and exit (myoglobin)
      2. Closing active site (carboxypeptidase)

   (c) Loop Motions

      1. Disorder-to-order transition covering active site (triose phosphate isomerase, penicillopepsin)
      2. Rearrangement as part of rigid-body motion (liver alcohol dehydrogenase)
      3. Disorder-to-order transition as part of enzyme activation (trypsinogen-trypsin)
      4. Disorder-to-order transition as part of virus formation (tobacco mosaic virus, tomato bush stunt virus)

   (d) Terminal Arm Motion

      1. Specificity of binding (λ-repressor-operator interaction)

II. *Rigid-Body Motions* (1 to 10 Å, 10^{-9} to 1 s)

   (a) Helix Motions

      1. Induction of larger-scale structural change (insulin)
      2. Transitions between substates (myoglobin)

   (b) Domains (hinge-bending) Motions

      1. Opening and closing of active-site region (hexokinase, liver alcohol dehydrogenase, l-arabinose binding protein)
      2. Increasing binding range of antigens (antibodies)

   (c) Subunit Motions

      1. Allosteric transitions that control binding and activity (hemoglobin, aspartate transcarbamoylase)

III. *Larger-Scale Motions* (>5 Å, 10^{-7} to 10^4 s)

   (a) Helix-coil transition

      1. Activation of hormones (glucagon)
      2. Protein folding transition

TABLE I—*Continued.*

(b) Dissociation/Association and Coupled Structural Changes

    1. Formation of viruses (tomato bushy stunt virus, tobacco mosaic virus)

    2. Activation of cell fusion protein (hemagglutinin)

(c) Opening and Distortional Fluctuations

    1. Binding and activity (calcium-binding proteins)

(d) Folding and Unfolding Transition

    1. Synthesis and degradation of proteins

---

folding transitions, in which the correlation of amplitude, energy, and time scale is expected to be approximately valid. However, in more localized events, often involving small displacements of a few atoms, the motion is slow because of a high activation barrier; examples are the aromatic ring flips in certain proteins. In this case the macroscopic rate can be very slow ($k \sim 1 \ s^{-1}$ at 300°K), not because an individual event is slow (a ring flip occurs in $\sim 10^{-12}$ s), but because the probability that a ring has sufficient energy to get over an activation barrier that is the order of 16 kcal/mol is very small ($\sim 10^{-12}$).

In Table I we list various types of motions with their length and time scales and give specific examples as to where such motions are found to have functional roles. It can be seen that for all three of the somewhat arbitrary classes (local, rigid body, and larger-scale motions) the listed time scales vary over many orders of magnitude. This is due primarily, as already suggested, to the presence of activation barriers that can slow down even the simplest motion (e.g., atomic fluctuations in a double-well potential). What the table makes clear is that a great range of motional phenomena are found to have a functional role. In some cases (e.g., the atomic fluctuations required for larger "rigid-body" displacements, the sidechain oscillations that play a role in the entrance and exit of ligands in myoglobin, the allosteric transition in hemoglobin, the disorder-order transition in going from inactive trypsinogen to the active enzyme trypsin) there exist detailed theoretical and/or experimental studies of the motions involved. In many other cases, the role of the motion has been inferred only from structural studies that show two or more different conformations.

The richness of the motional phenomena that are involved in protein function, which is only hinted at in Table I, makes the field of macromolecular dynamics one of the most exciting and rapidly developing areas of chemical physics. It is our hope that the reader will come away from this volume with an understanding of the nature of protein motions, their functional role, and the methods used for studying them.

CHAPTER III

POTENTIAL FUNCTIONS

In this and the two following chapters we present the methodological basis of theoretical studies of the internal motions and of the thermodynamics of proteins and other macromolecules of biological interest. The first step is to construct a potential energy surface, the energy of the system as a function of the atomic coordinates. The potential energy can be used directly to determine the relative stabilities of the different possible structures of the system. To obtain the forces acting on the atoms of the system, the first derivatives of the potential with respect to the atom positions are calculated. These forces can be used to determine dynamical properties of the system; e.g., by solving Newton's equations of motion to describe how the atomic positions change with respect to time. From the second derivatives of the potential surface, the force constants for small displacements can be evaluated and used to find the normal modes. The normal modes provide an alternative approach to the dynamics in the harmonic limit.

In this chapter we outline the nature of the potential functions (force fields) that are generally employed for macromolecules. In the next chapter we describe methods that can be used to determine the dynamics over the wide range of time scales that is of interest. In the final chapter on methodology we present some techniques for evaluating thermodynamic properties; dynamical calculations are shown to play an important role in many of these techniques.

A. THEORETICAL BASIS

In considering the method to use for determining the potential surface, the most direct approach would be to solve the quantum-mechanical problem in the Born-Oppenheimer approximation for the system of interest. However, it is necessary to go beyond this type of approach for two reasons. The first is that the number of electrons and nuclei in the system is an order of magnitude larger than the number of atoms. Since the number of atoms can be quite large (e.g., it is not uncommon in the study of proteins in aqueous solution to have to treat several thousand atoms), reducing the number of degrees of freedom without altering the essential properties of the system is important.

Moreover, solving the quantum-mechanical problem even for a small number of atoms is exceedingly time consuming. Quantum-mechanical calculations that are of useful accuracy can be made for 20 or so atoms at a few geometries. Thus quantum-mechanical methods in the present context are limited to small-fragment studies (e.g., model system reactions and component surfaces required for empirical energy functions). The second difficulty with quantum-mechanical approaches is that their accuracy is generally not sufficient for the types of interaction energies that are required. In the best calculations for simple hydrides (e.g., CH_4) there is an uncertainty in the atomization energy of about 2 kcal/mol.[57] As soon as there are two heavy atoms (i.e., atoms other than hydrogen), the results that can be achieved are considerably worse. For most of the properties of the macromolecular systems of interest here, any quantum-mechanical calculation that is feasible is not sufficiently accurate.

To obtain potential energy surfaces for proteins with the required accuracy and speed, it is necessary to introduce a simpler model which is calibrated by fitting it to experimental or quantum mechanical information. When working with macromolecules, there is a need to have available a reliable method for calculating interaction energies many times (10^4 to 10^6 energy calculations) for systems of hundreds to thousands of atoms. Such a method is supplied by empirical energy functions. However, there is a price to pay for introducing such a model into the calculation. Empirical energy functions do not have the generality of quantum mechanical calculations; they are at best limited to the systems for which they were designed. Moreover, although the quantum-mechanical approach involves only pairwise-additive interactions, the empirical expressions include many-body interactions.

The effective (Born-Oppenheimer) forces describing the atomic interaction may be either attractive or repulsive. The attractive contributions, which tend to vary slowly with distance, are important in determining the global cohesive properties of a system and can be important in stabilizing specific structures. The repulsions, which have a much steeper distance dependence, play the primary role in determining the allowed conformations and the dynamics. In a certain sense, the van der Waals repulsions are the most important interactions between nonbonded atoms, as they give rise to the finite size of the atoms. Hard (impenetrable)-sphere nonbonded interactions are among the simplest repulsive models for atoms and have been used in a wide variety of problems in chemical physics. Such a simple model[58] was employed in 1963 to predict the allowed conformations of amino acids, the basic building block of proteins. Fused hard-sphere models have also been used[59] for describing the configurational distributions and free energies of alkane chains in the liquid state. With relatively little complication it is possible to introduce considerably more realistic models for molecules that consist of soft-sphere atoms with electric multipole moments connected by "springs" that account for the local covalent structure. These elements are the essence of most empirical energy functions.

B. FORM OF POTENTIAL FUNCTIONS

To obtain the accuracy required for a realistic analysis of the structure and dynamics of macromolecules it is necessary to use a relatively complex form for the empirical potential function and to optimize the values of the parameters that determine the magnitudes of the different contributing terms. In general, the function will have terms that depend not only on the relative position of all pairs of atoms but certain triples and quadruples of atoms as well. Usually, one does not need to go beyond four-body terms in the model potential function. This approach to calculating energies is often referred to as molecular mechanics.[60,61]

A number of potential functions for use with polyatomic systems have become available in recent years.[62-68] Features common to most of them include a harmonic restoring force between bonded nearest neighbors, a penalty for deforming the angle between three neighboring atoms, a dihedral torsional potential to allow for the hindered rotation of groups about a bond, and nonbonded interactions between separated atoms. Two separate molecules interact with each other only through the nonbonded interactions unless reactions are being considered. Thus nonbonded interactions are common to both intermolecular and intramolecular potentials, whereas the other terms are strictly intramolecular.

The important nonbonded interactions between atoms consist of a part that accounts for the excluded volume, the repulsive van der Waals term already mentioned, a part that models the dispersion attraction, and a part that characterizes the electrostatic interaction due to the partial charge and higher electric moments associated with each atom. The repulsive forces arise from a combination of internuclear repulsions and the Pauli exclusion principle. These forces for systems at or near standard temperatures and pressures may be modeled by rather simple, positive definite forms. At extreme pressures or temperatures, not of general interest for biological macromolecules, rather more sophisticated expressions may be needed.[69] The dispersion force or London force arises from small fluctuations of the charge distribution of an atom in the presence of another atom. These fluctuations give rise to an attractive dipole-dipole interaction that London first showed decreases with the inverse sixth power of the separation distance.[70] This relatively short-ranged attraction and the even-shorter-ranged repulsion are usually considered together in formulating the potential energy function. Some of the more common representations employ a combination[60,61] of an exponential repulsion with an inverse sixth-power attraction (exp-6) or an inverse twelfth-power repulsion with the inverse sixth-power attraction (6-12), the widely used Lennard-Jones potential form.[71] For certain cases softer repulsive exponents, usually an inverse tenth or ninth power, have been suggested.[61,63] Also, higher-order attractive dispersion force terms have been included.[72]

The other contribution to the nonbonded interactions arises from the effective partial charges that reside on the atoms. When atoms of differing electronegativity are chemically bonded to each other, net electronic charge tends to flow from the less electronegative atom to the more electronegative atom until a balance is achieved. This rearrangement of the charge distribution along the chemical bonds is represented by assigning partial charges (or higher moments) to the atoms. In addition, there are groups present in proteins (e.g., aspartic acid, arginine), in lipids (e.g., the head groups), and in nucleic acids (e.g., the phosphate groups) that are ionized at physiological pH and therefore carry a net charge. Charge-charge attractions resulting from opposite partial or net charges tend to be larger than other nonbonded attractions. Also, they are of considerably longer range than the dispersion forces since they decrease as the inverse first power of the distance. In some potential functions these terms account for the formation of hydrogen bonds and salt links; in others, the hydrogen bonds are treated by the introduction of special terms.[62,65]

The sign and magnitude of the net atomic partial charges may be determined by a variety of methods. The most widely used approaches are based on an analysis of the charge distribution as given by *ab initio* calculations of the ground-state electronic charge density. The simplest and most straightforward method makes use of a Mulliken[73] or Lowdin[74] population analysis. However, the results tend to be basis-set dependent and do not necessarily reflect the full set of electrostatic moments calculated from the electron density. The lowest-order moments estimated from such population analyses have been used in a number of molecular models.[65,66] These estimates for the charges are often refined by comparison of calculated quantities with appropriate experimental data (e.g., the interaction energy of two polar molecules). This has been particularly effective in the case of models developed for both solid-state and liquid-state simulations.[75] It is also possible to use an electronic wave function directly to calculate the electrostatic potential on a grid of points at a specified set of distances from the molecule.[76] A point-charge (or higher-order) model can then be fitted to reproduce the electrostatic potential.

The above discussion neglects polarization effects. In part, these are taken into account by the empirical scaling procedures (e.g., the dipole moment of the water molecule in condensed-phase models is chosen to be larger than the gas-phase value).[12] Internal charge rearrangements due to conformational changes are generally assumed to be negligible; they have been explored in some molecular mechanics models[77] and may be required for very accurate results.

The simplest and most widely used expression for the nonbonded interactions is of the Lennard-Jones plus Coulomb or 1-6-12 form; i.e.,

$$V_{nb} = \sum_{\substack{\text{nonbonded} \\ i,j \text{ pairs}}} 4\epsilon_{ij}\left[\left(\frac{\sigma_{ij}}{r}\right)^{12} - \left(\frac{\sigma_{ij}}{r}\right)^{6}\right] + \frac{q_i q_j}{\epsilon r} \tag{1}$$

where r, ϵ_{ij}, σ_{ij}, q_i, and ϵ are the nonbonded distance, the dispersion well depth, the Lennard-Jones diameter, the charge, and a dielectric parameter, respectively. In calculating the interactions between molecules, the terms between all atom pairs are counted. However, for intramolecular interactions, only those between atoms separated by at least three (sometimes four) bonds are included. This is to take account of the fact that van der Waals spheres overlap considerably at the chemical bonding distance and that the chemical bonding interactions are described by separate terms in the potential function (see below).

If all of the atoms and charges in the system of interest are explicitly represented and atomic polarization is included, the use of a dielectric constant other than unity would be inappropriate. A variety of models has been used, however, to approximate the dielectric behavior of a macromolecular system where the solvent was not explicitly included. Dielectric constants for the protein interior between 2 and 10 have been employed, as has a distance-dependent dielectric response equal to the distance in angstroms.[78] Also, simple forms of the Kirkwood-Westheimer-Tanford model[79] have been used to approximate the effect of the aqueous solvent. An approach that may improve our understanding in this area employs linear response theory to evaluate the spatially dependent dielectric response.[80] In any such model it is necessary to consider the frequency dependence of the dielectric constant relative to the time scale of the dynamic process under consideration.

Bonds are frequently modeled by a simple harmonic potential form, although Morse functions[81] and other more complex expressions have been used, particularly for relatively simple molecules.[61] This is appropriate because most of the motions that occur in proteins at ordinary temperatures leave the bond lengths (and bond angles) near their equilibrium values, which appear not to vary by large amounts throughout the molecule (e.g., the standard dimensions of the peptide group first proposed by Pauling[82] provide an accurate representation). For bonded atoms, the so-called 1,2 pairs, a harmonic interaction potential of the form

$$V_{\text{bond}} = \sum_{\substack{1,2 \\ \text{pairs}}} \tfrac{1}{2} K_b (b - b_0)^2 \tag{2}$$

is used, where b, K_b, and b_0 are the bond length, the bond-stretching force constant, and the equilibrium distance parameter, respectively.

The terms that keep the bond angles near the equilibrium geometry involve triples of atoms or neighboring pairs of bonds. Frequently, a harmonic force dependence on the angle is sufficient and is taken to have the form

$$V_{\text{bond angle}} = \sum_{\substack{\text{bond} \\ \text{angles}}} \tfrac{1}{2} K_\theta (\theta - \theta_0)^2 \tag{3}$$

where θ, K_θ, and θ_0 are the bond angle, the angle bending force constant, and the equilibrium value parameter, respectively. Particularly for bond angles involving hydrogen atoms, it has been found that a direct interaction between the first and third atom defining the angle is helpful in fitting the observed vibrational spectra. A functional form to treat this case, the so-called Urey and Bradley interaction,[83] consists of a shifted harmonic oscillator

$$V_{\text{U-B}} = \sum_{\substack{1,3 \\ \text{pairs}}} \tfrac{1}{2} K_{ub} (S - S_0)^2 + \sum_{\substack{1,3 \\ \text{pairs}}} K'_{ub} (S - S_0) \tag{4}$$

where S, K_{ub}, K'_{ub}, and S_0 are the 1,3 distance, the harmonic Urey-Bradley force constant, the linear Urey-Bradley force constant, and the 1,3 equilibrium distance, respectively. One can algebraically reduce the set of three constants, K_{ub}, K'_{ub}, and S_0 to two without loss of generality.[84] Although the Urey-Bradley parameters tend to be highly correlated with the parameters for the valence angle,[78] good results have been obtained by using a combination of bond angle and Urey-Bradley terms.[78,85]

It has been found[60] that the hindered rotation about single and partial double bonds cannot be modeled with sufficient accuracy by the terms that have been discussed so far. If the Lennard-Jones parameters are made large enough to ensure the proper barrier for the torsional motion, they no longer provide a good representation of the intermolecular interactions. Thus it is usually necessary to introduce an explicit torsional potential with the proper rotational symmetry and energy barriers. A commonly used technique is to use a cosine expansion and keep only the lowest-order term for each torsion. The sum over all the torsional degrees of freedom for a large system is then

$$V_{\text{torsion}} = \sum_{\substack{\text{dihedral} \\ \text{angles}}} K_\phi [1 + \cos(n\phi - \delta)] \tag{5}$$

where ϕ, K_ϕ, n, and δ are a dihedral angle, its force constant, multiplicity, and phase, respectively. With this form one can parameterize the common twofold barriers, such as occur in the peptide bond of an amide, and the threefold barriers, such as those encountered in hydrocarbon chains. For some molecules (e.g., sugars), additional terms in the torsional expansion have to be included.[67,68]

A typical empirical potential energy surface is the sum of all the interactions described above. It can be written schematically in the form

$$V_T = \sum_{\substack{1,2 \\ \text{pairs}}} \tfrac{1}{2}K_b(b - b_0)^2 + \sum_{\substack{\text{bond} \\ \text{angles}}} \tfrac{1}{2}K_\theta(\theta - \theta_0)^2$$

$$+ \sum_{\substack{\text{dihedral} \\ \text{angles}}} K_\phi[1 + \cos(n\phi - \delta)]$$

$$+ \sum_{\substack{\text{nonbonded} \\ i,j \text{ pairs}}} 4\epsilon_{ij}\left[\left(\frac{\sigma_{ij}}{r}\right)^{12} - \left(\frac{\sigma_{ij}}{r}\right)^6\right] + \frac{q_i q_j}{\epsilon r} \tag{6}$$

This form of potential function is widely used because it appears to be a satisfactory compromise between simplicity and accuracy.[65,78] It should be mentioned, however, that additional terms have been introduced in some cases. Certain of these have already been mentioned. Others include cross terms between some of the interactions, usually to obtain better fits of vibrational spectra. An example would be a coupling between the dihedral-angle and bond-angle potential terms.[86]

Simplifications have been introduced in some applications to reduce the number of atoms and therefore the number of degrees of freedom that have to be included explicitly. One reduced set is that in which all hydrogen atoms are included as part of the heavy atom to which they are attached. This "extended atom" representation, which has a long history in scattering and liquid-state problems (e.g., two methanes treated as spheres with suitably chosen van der Waals parameters), was introduced in early energy minimization and dynamics calculations for macromolecules.[62,65] At present, one of the most common approaches is to include only nonpolar hydrogens as part of extended atoms (essentially only CH_3, CH_2, and CH groups) and to treat all other (polar) hydrogens (NH, OH, etc.) explicitly. Such an approach, with explicit polar hydrogens, appears to be satisfactory for many purposes; it is particularly well suited for the accurate treatment of the important hydrogen-bonding interactions. Potential functions that include all hydrogens are being employed for problems where the detailed behavior of the nonpolar hydrogens is significant (e.g., certain binding interactions, analysis of NMR and vibrational spectroscopic experiments).[62-67]

C. PARAMETER DETERMINATION

Given the form for the potential function, it is necessary to determine the parameters corresponding to the set of terms appearing in the energy expression (e.g., Eq. 6) so as to fully characterize the potential energy surface for the system of interest. In general, an initial guess is made for the parameters by use of data from model systems. This guess is refined by trial and error and via nonlinear least-squares techniques making use of a variety of information for molecules related to the system one wishes to simulate. What makes such a parameter evaluation feasible for biological macromolecules is that the number of different atoms involved is small (i.e., H, C, N, O, S, P plus specific metal atoms) and the types of bonding (with the associated energy terms) are limited; e.g., there are only 20 or so amino acids that occur in proteins and the number of different nucleic acid bases, lipids, and sugars is small. However, this also points to a limitation of such empirical potential functions. If different types of molecules are of interest, such as might be synthesized to obtain a more potent drug, inhibitor, or antibiotic analogue, new terms and the associated parameters may have to be introduced into the potential function.

The Lennard-Jones parameters are obtained from a number of sources,[87] such as viscosity data, scattering data, and high-level quantum-chemical calculations. Other types of information that can be used to refine these estimates are crystal structures[62] and liquid structure data.[75] Methods for determination of the atomic partial charges have already been discussed. Dihedral angle multiplicities, bond lengths, and bond angles may be evaluated by use of quantum chemical calculations for simple model systems, experimental structural information, and chemical intuition. The force constants for the individual terms may then be obtained from vibrational and distortional data. One of the best approaches is to determine the minimum energy structure, calculate the normal modes,[68] and compare them with assigned vibrational spectra.[88] The force constants (and other parameters) are then least-squares adjusted until the normal-mode frequencies and eigenvectors match the experimental results. Often it is desirable to attempt such fits with a variety of weighting factors in the least-squares procedure to assure that a well-balanced parameterization results.

Experience has shown that useful information can be obtained by employing the presently available functional forms and parameter sets. Illustrations of the applications to dynamic and thermodynamic simulations are given in subsequent chapters. However, it must always be remembered that whatever the precision of the motional or thermodynamic properties obtained from the simulation methods, the ultimate accuracy of the results depends on the potential function that was used. Approximations or errors present in the potential are reflected in the applications. This implies that caution must be exercised in the choice of a potential function and its parameterization to make certain that it is suitable for the systems and questions under investigation.

CHAPTER IV

DYNAMICAL SIMULATION METHODS

The simulation methodologies described here are designed to study structural, dynamical, and thermodynamic properties of biological macromolecules. The most exact and detailed information is provided by molecular dynamics simulations, in which one uses a computer to solve the Newtonian equations of motion for the atoms of the macromolecules and, in principle, the surrounding solvent. The early molecular dynamics applications were limited to biological molecules in vacuo; i.e., the simulations were carried out in the absence of any solvent or in the presence of only a few of the "crystal waters." This made it possible to simulate the dynamics of small proteins (up to a 1000 atoms) for up to a few hundred picoseconds. Such periods are long enough to characterize completely the librations of small groups in the protein and to determine the dominant contributions to the atomic fluctuations. The advent of new supercomputer technologies and recent theoretical developments are making it possible to study biomolecules in solution and to extend the simulations to times on the order of nanoseconds. To examine still slower and more complex processes, it is necessary to use methods other than straightforward molecular dynamics simulations. These include harmonic dynamics, stochastic dynamics, and activated dynamics methods, each of which is particularly useful for certain types of problems. In this section we review these methods, all of which are currently in use for simulating the dynamics of macromolecules of biological interest.

A. GENERAL FEATURES OF MOLECULAR DYNAMICS METHODS

In a molecular dynamics simulation the classical equations of motion for the solute (biopolymer) and solvent atoms, if treated explicitly, are integrated numerically; i.e., Newton's equations of motion,

$$m_i \frac{d^2 \mathbf{r}_i}{dt^2} = -\nabla_i [U(\mathbf{r}_1, \mathbf{r}_2, \ldots, \mathbf{r}_N)] \qquad i = 1, N \qquad (7)$$

are solved to obtain the atomic positions and velocities as a function of time;[12] here m_i and $\mathbf{r}_i$ represent the mass and position of particle i and $U(\mathbf{r}_1, \mathbf{r}_2, \ldots, \mathbf{r}_N)$ is the potential energy surface that depends on the positions of the N particles in the system. For simulations involving biopolymers such as proteins and nucleic acids, the initial positions for the atoms of the biopolymer are obtained from a known X-ray structure; the positions of the solvent atoms are usually determined by fitting the biomolecule into a preequilibrated box of solvent atoms. The X-ray structure, and the solvent when present, are first refined using an energy minimization algorithm (see Chapt. IV.H).[65] This energy refinement relieves local stresses due to nonbonded overlaps, as well as bond-length and bond-angle distortions in the X-ray structure and the surrounding solvent. The atoms are then assigned velocities from a Maxwellian distribution at a temperature below the desired temperature; usually, a temperature near zero is used as the starting temperature. The system is then equilibrated by integrating the equations of motion while adjusting the temperature and density to the appropriate values. The temperature is brought into the range of interest by incrementally increasing the velocities of all the atoms, either by reassignment from a Maxwellian distribution at an increased temperature or by scaling all velocities. The temperature, $T(t)$, at any given time t is defined in terms of the mean kinetic energy by

$$T(t) = \frac{1}{(3N - n)k_B} \sum_{i=1}^{N} m_i |\mathbf{v}_i|^2 \tag{8}$$

where $(3N - n)$ is the total number of unconstrained degrees of freedom in the system, $\mathbf{v}_i$ is the velocity of atom i at time t, and k_B is the Boltzmann constant. It is clear from this expression that scaling the velocities by a factor of $\sqrt{T'/T(t)}$ will result in a mean kinetic energy corresponding to a temperature of T'. The heating and equilibration process decreases the probability that localized fluctuations in the energy (e.g., "hot spots") will persist throughout the simulation. Following the initial equilibration/thermalization period (on the order of 10 to 50 ps are required), the system is further equilibrated for some period of time (10 to 25 ps), during which no adjustments are made. Once the properties of the system are stable (e.g., the average kinetic energy remains constant), the trajectory is calculated for an extended period to be used for analysis. To date, the available simulations of protein and nucleic acid systems range in length from ~25 ps to 1 ns. However, it is clear from studies in the area of molecular fluids, where simulations of several nanoseconds[89] exist, that corresponding extensions of the time scale will be possible, and sometimes necessary, in the area of macromolecules.

Several algorithms for integrating the equations of motion in Cartesian coordinates are used in protein dynamics calculations. Most common is the Verlet algorithm,[90] which is widely employed in statistical mechanical simulations. Also, the Gear predictor-corrector algorithm,[91] which is commonly used in small molecule trajectory calculations, has been applied to macromolecules.[15] The specific algorithms are presented in a subsequent section (Chapt. IV.G.).

Given a trajectory, all of the equilibrium and dynamical properties of the system can be determined, in principle. In the case of average quantities it is necessary to ensure that a representative system is used for computing the averages since one is replacing an ensemble average by a time average over a single system. This has been achieved for systems where strong coupling among the various degrees of freedom leads to rapid equilibration (e.g., in the simulation of simple fluids).[11] For a single macromolecule in solution, obtaining representative results is more difficult. Attempts to do so are based on two different approaches. One method is to use a single long trajectory. This technique is expected to be most accurate when energy redistribution is rapid; proteins appear to have fairly fast energy redistribution. For systems with a nearly harmonic potential function (e.g., small peptides), the method of choice is to average over a series of trajectories with different initial conditions.[84] When neither a long trajectory nor multiple trajectories are available, caution must be exercised in interpreting the results; a simulation of the average motions of one member of an ensemble does not necessarily correspond to the ensemble average, although it may, in fact, be representative. It should also be noted that taking derivatives tends to amplify errors, while integration suppresses statistical errors. Thus properties such as the heat capacity or compressibility are expected to be less accurate than the average quantities from which they are derived.

All of the simulation approaches, other than harmonic dynamics, include the basic elements that we have outlined. They differ in the equations of motion that are solved (Newton's equations, Langevin equations, etc.), the specific treatment of the solvent, and/or the procedures used to take account of the time scale associated with a particular process of interest (molecular dynamics, activated dynamics, etc.). For example, the first application of molecular dynamics to proteins considered the molecule in vacuum.[15] These calculations, while ignoring solvent effects, provided key insights into the important role of flexibility in biological function. Many of the results described in Chapts. VI–VIII were obtained from such vacuum simulations. Because of the importance of the solvent to the structure and other properties of biomolecules, much effort is now concentrated on systems in which the macromolecule is surrounded by solvent or other many-body environments, such as a crystal.

B. MOLECULAR DYNAMICS WITH CONVENTIONAL PERIODIC
BOUNDARY CONDITIONS

One method of reducing the complications inherent in the study of solvated molecules is to impose periodic boundaries on the central cell of molecules whose dynamics are to be considered explicitly. This cell contains the molecule or molecules of interest, together with an appropriate number of solvent molecules. The central cell is generally cubic or a rectangular parallelepiped, but it may also be a truncated octahedron or have a more general geometry.[92,93] It is surrounded by periodic images of itself. The images are defined by transformations related to the symmetry of the central cell (see Fig. 7) and the particles in the image cells undergo the same motions as those of their partners in the central cell. A dynamics simulation is carried out for the atoms in the central cell in the force field of the image cells. The resulting trajectory thus corresponds to that for an infinite periodic system.

Dynamics simulations employing conventional periodic boundary conditions are usually carried out in the microcanonical (constant N, constant volume, constant energy) ensemble. However, techniques have been introduced that allow NVT (constant N, constant volume, constant temperature), NPH

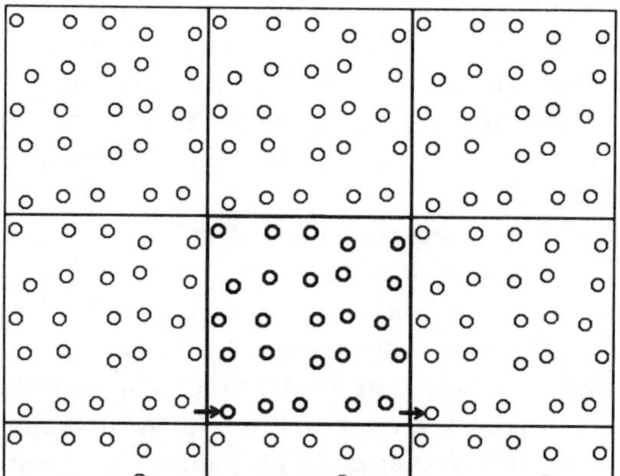

Figure 7. Periodic boundaries in two dimensions. Illustration of a primary cell (bold) and its eight nearest-neighbor images. The arrows indicate how periodicity is enforced; i.e., a particle leaves on the right and reenters on the left.

(constant N, constant pressure, constant enthalpy), and NPT (constant N, constant pressure, constant temperature) ensembles to be treated.[94-96] These methods rely on coupling the central cell of atoms to a constant-temperature bath and/or a constant-pressure "piston." For example, in one approach which is described as "weak coupling to an external bath," constant-temperature conditions are imposed by adding Langevin dissipative forces to *all* atoms of the central cell;[96] i.e., the equations of motion have the form

$$m_i \frac{d^2 r(t)}{dt^2} = -\nabla_i [U(\mathbf{r}_1, \mathbf{r}_2, \ldots, \mathbf{r}_N)] + m_i \beta \left[\frac{T_0}{T(t)} - 1 \right] \frac{dr_i(t)}{dt}$$

$$i = 1, N \quad (9)$$

In this equation, β is an arbitrary frictional drag parameter (inverse time constant), chosen as the coupling parameter which determines the time scale of temperature fluctuations, T_0 is the mean temperature, and $T(t)$ is the temperature at time t (see Eq. 8). Constant-pressure conditions are enforced with a proportional scaling of all coordinates and the box length by a factor related to the isothermal compressibility for the system.[94] In principle, simulations carried out in all of these ensembles should yield the same results for *equilibrium* properties in systems of sufficient size; of course, when differing ensembles are employed, appropriate corrections (e.g., a PV correction to compare the NVT and NPT ensembles) must be introduced. So far, little work has been done to determine quantitatively the system size needed to reach the "thermodynamic" limit.

Molecular dynamics with periodic boundary conditions is presently the most widely used approach for studying the equilibrium and dynamic properties of pure bulk solvent,[97] as well as solvated systems. However, periodic boundary conditions have their limitations. They introduce errors in the time development of equilibrium properties for times greater than that required for a sound wave to traverse the central cell. This is because the periodicity of information flow across the boundaries interferes with the time development of other processes. The velocity of sound through water at a density of 1 g/cm^3 and 300 K is ~15 Å/ps; for a cubic cell with a dimension of 45 Å, the cycle time is only 3 ps and the time development of all properties beyond this time may be affected. Also, conventional periodic boundary methods are of less use for studies of chemical reactions involving enzyme and substrate molecules because there is no means for such a system to relax back to thermal equilibrium. This is not the case when alternative ensembles of the constant-temperature variety are employed. However, in these models it is not clear that the somewhat arbitrary coupling to a constant temperature heat bath does not influence the rate of "reequilibration" from a thermally perturbed state; the mechanism of relaxation is not correctly represented by the available constant-temperature algorithms. For such nonequilibrium problems, alternative simulation methods may be more appropriate.[98] Additional complications arise in the use of periodic boundary conditions to simulate very dilute solutions of biomolecules. In essence, the concentration of the solution being studied is dictated by the number of solvent molecules present in the central cell; e.g., to study a $0.01M$ solution of NaCl, including the ion-ion interactions, would require about 25 molecules of NaCl and about 137,000 water molecules. This problem is less important when the range of the atom interaction potentials is short enough (i.e., for nonionic systems) to allow for a central cell which is sufficiently large that the solute molecule never sees its own image. However, the extent to which solvent-mediated interactions correlate the solute particles in a manner characteristic of more concentrated solutions is not clear.

C. MOLECULAR DYNAMICS WITH STOCHASTIC BOUNDARY CONDITIONS

In many cases the processes of interest (e.g., energy transport and chemical reactivity in biomolecules) occur in a localized region of the protein-solvent system. Examples of biochemical processes for which this is likely to be true are enzyme reactions and ligand binding for transport and storage, as in myoglobin, where structural and mechanistic studies suggest that biological activity is linked to the dynamics occurring in the neighborhood of the active or binding site. Another important case is the thermodynamic change resulting from localized structural perturbations due to mutation of a protein or binding of a substrate to an enzyme (see Chapt. X). Conventional molecular dynamics techniques may be an inefficient, and in some cases an inappropriate, way of studying the essential dynamics of such systems. Special methodologies that eliminate the uninteresting motions and focus on a specific spatially localized region of the biomolecular system appear best suited for such problems. Further, for reactions it is important to provide a realistic mechanism for thermal equilibration of the system.

The stochastic boundary approach in conjunction with molecular dynamics is an approximate technique for studying such localized events in many-body systems.[99] The method was developed initially to study nonequilibrium phenomena[98] (e.g., chemical reactions and atomic diffusion across thermal gradients) and hence is well suited for some of the problems of interest here. The approach has been used to treat simple fluids,[100,101] as well as more complex fluids, including water,[102] and solvated biomolecules.[103]

An essential feature of the stochastic boundary methods is a partitioning of the many-body system into several regions. The regions are delineated based on their spatial disposition with respect to a primary area of interest. The entire system is divided into a "reaction" zone and a reservoir region. The reaction zone contains the portion of the system that is of interest, and the reservoir region is the portion of the system that does not participate directly. This partitioning is analogous to the division of many-body systems that has been used in other applications of nonequilibrium statistical mechanics (e.g., generalized Langevin theory) to study reaction and atomic dynamics in condensed phases and on surfaces.[104,105] The reservoir region is excluded from the calculation and its effect is replaced by appropriately chosen mean and stochastic forces. To introduce these in the stochastic boundary methodology, the reaction zone is further divided into a reaction region and a buffer region, with stochastic forces applied to atoms in the buffer region. In this manner, buffer region atoms act as a heat bath for thermal fluctuations occurring in the reaction region. This decomposition is similar to the expansion of the heat bath degrees of freedom in terms of the chain representations employed in generalized Langevin theory.[106,107] The difference is that the primary zone, in the terminology of generalized Langevin equation theory, is much larger and more complicated in the systems for which the stochastic boundary approach was designed; i.e., many of the "heat bath" effects, which would have to be accounted for in the stochastic contributions to the generalized Langevin heat bath forces, are included explicitly in the reaction zone of the stochastic boundary simulation. Hence the stochastic heat bath forces may be assumed to have a relatively simple form; specifically, simple Langevin dissipative and random forces are used. The reaction region atoms are treated by conventional molecular dynamics (i.e., via Newton's equations) and are not directly coupled to a Langevin heat bath as they are in the constant-temperature molecular dynamics algorithms described above.

The basic partitioning is illustrated schematically in Fig. 8a and realistically in Fig. 8b for a simulation study focusing on the dynamics of a tryptophan ring in the protein lysozyme.[108] With the division indicated in the figure the total number of atoms to be simulated is 696 (294 protein atoms and 134 water molecules). This is a great reduction from the estimated 11,766 atoms (1266 protein atoms and 3500 water molecules) that would be necessary if conventional periodic boundary conditions were employed; the estimate is based on using a 50-Å cubic cell, a 26-Å sphere to represent lysozyme, and 1 g/cm^3 density for water.

The stochastic boundary methodology requires a scheme for partitioning the protein-solvent system and a procedure for calculating the mean (boundary) forces, as well as the appropriate simulation equations for the various regions. The partitioning for each specific system is expected to be somewhat different. However, a few general rules can be stated. Initially, one defines the geometric center of an "active site," the region of primary focus, and parti-

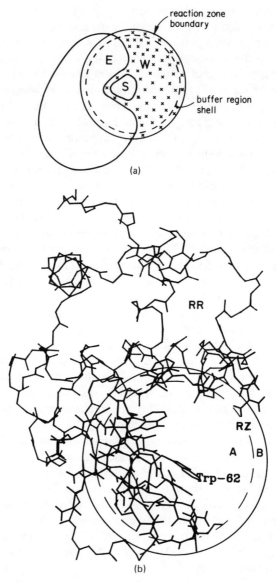

Figure 8. (*a*) Schematic partitioning of an enzyme (E)–substrate (S)–water (W) system into a spherical zone and surroundings. (*b*) Partitioning of lysozyme in stochastic boundary molecular dynamics simulation of the active site. The entire molecule is partitioned into a reaction zone (RZ) and a reservoir region (RR); the reservoir region (in the RR only mainchain atoms are depicted) is neglected. The reaction zone is further divided into a reaction region (A) and a buffer region (B). The partitioning is with respect to the center of the active site, the geometric center of the Trp-62 sidechain, and the dynamics is performed only on the reaction zone atoms.

tions the system into approximately spherical layers centered on this point. The partitioning separates the biomolecule(s) and solvent into two regions, labeled RZ (reaction zone) and RR (reservoir region) in Fig. 8. The extent of the RZ is such that most nonbonded interactions of atoms in the RR with "active-site" atoms would be negligible; values between 9 and 20 Å have been used as the radius of the reaction zone. The criterion used to determine which atoms near the boundary are included in the simulation is based on residues (a residue here means an amino acid residue for proteins or an entire base for nucleic acids); i.e., the entire residue is included if *any* atom of that residue is inside the spherical RZ.

A second stage of partitioning involves separating the system into a reaction region, labeled A in Fig. 8a, and a buffer region, labeled B. The buffer region atoms are labeled as those atoms with a separation greater than R from the center of the reaction zone. This labeling is a dynamic one since groups may diffuse across this boundary; e.g., for the solvent the buffer region atom labeling is updated during the course of the dynamics (e.g., every 20 steps). The buffer region atoms interact with a stochastic heat bath, via random fluctuating forces and dissipative forces, that account for the dynamical character of the neglected reservoir region atoms.

To provide an efficient simulation algorithm, the heat bath forces are assumed to be simple. They are represented as Langevin dissipative forces, proportional to the atomic velocities,

$$\mathbf{F}_d(t) = -m_i \beta \mathbf{v}_i(t) \qquad (10)$$

and Langevin random forces, $m_i\mathbf{f}$, which satisfy

$$\langle \mathbf{f}(t) \rangle = 0 \quad \text{and} \quad \langle \mathbf{f}(t) \cdot \mathbf{f}(0) \rangle = 6k_B T \beta \delta(t) \qquad (11)$$

The proportionality constant, β, in the expressions above is the friction coefficient. It is obtained from the inverse of the velocity correlation function relaxation time. Typically, values ranging from 50 to 200 ps^{-1} have been employed.[101-103]

To account for the neglected average interaction with the reservoir, static boundary forces are also applied to the system. The range of these forces is, in general, governed by the extent of interparticle shielding, i.e., the distance of a given atom from the reservoir region. The explicit form of the mean force is rigorously governed by the many-body distribution functions for the system. These quantities are very complicated and their calculation through statistical mechanical relationships is, in most cases, very difficult.[109] However, for homogeneous fluids one can introduce a satisfactory analytic approximation to this force. For protein atoms the choice of the boundary force is based on empirical considerations.

In the case of solvent molecules within RZ, the aim is to calculate the average force on a molecule, at r_0 inside RZ, from molecules in RR. This force may be represented by the mean-field force arising from an equilibrium distribution of solvent outside RZ.[101] There results the expression

$$\mathbf{F}_B(\mathbf{r}_0) = \int_{>RZ} d\mathbf{r}_T \mathbf{F}(\mathbf{r}_0 - \mathbf{r}_T) \rho_T g(\mathbf{r}_0 - \mathbf{r}_T) \tag{12}$$

In Eq. 12, $\mathbf{F}_B$ is the boundary force at $\mathbf{r}_0$, $\mathbf{F}(\mathbf{r}_0 - \mathbf{r}_T)$ is the force of interaction between a particle at $\mathbf{r}_T$ in RR and a particle at $\mathbf{r}_0$ in RZ, and $d\mathbf{r}_T \rho_T g(\mathbf{r}_0 - \mathbf{r}_T)$ is the probability of the pair $(0, T)$ having a separation $\mathbf{r}_0 - \mathbf{r}_T$. The boundary force may be written as the gradient of a potential, the boundary potential. The boundary potential for the oxygen atom of ST2 water[12] in an 11 Å reaction zone is plotted in Fig. 9. In the calculation of this potential only the van der Waals part of the ST2-ST2 interaction was included in Eq. 12. A methodology that consistently incorporates electrostatic forces into the boundary potential is under development.[110] In its present simplified form, the model has proven successful in the simulation of localized regions of pure

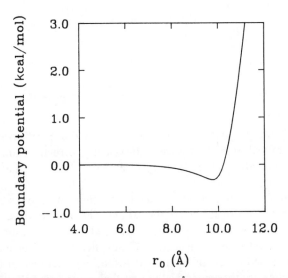

Figure 9. Boundary potential (kcal/mol) versus r_0 (Å) for the oxygen atom of an ST2 model water molecule. The calculation is for an 11-Å reaction zone and only the van der Waals forces are included (see Eq. 12).

ST2[102] and TIP4P[111] water models. Somewhat artificial orientation effects do occur near the boundary,[111] but they appear to be unimportant in describing the local dynamics in RZ away from the boundary.

Boundary forces for the protein are calculated from the known mean-square fluctuations of the atoms in the macromolecule. The difference between the choice of boundary force for liquids as just described and that for a macromolecule comes from the fact that the latter has a well-defined average structure and that in many cases atoms only undergo localized motions relative to the average structure (see Chapt. VI). To take account of the localized nature of the atomic motions, harmonic restoring forces are imposed on the heavy atoms in the buffer region, region B in Fig. 8a. The result is that the buffer region protein atoms remain close to their average positions and aid in maintaining the structural integrity of the remainder of the molecule. The protein boundary forces are given by

$$\mathbf{F}_B(\mathbf{r}_i) = -m_i\Omega_i^2\Delta\mathbf{r}_i = -\frac{3k_bT}{\langle\Delta\mathbf{r}_i^2\rangle}S(\mathbf{r}_i)\Delta\mathbf{r}_i \qquad (13)$$

Here Ω_i^2 is the force constant for atom i and $\langle\Delta\mathbf{r}_i^2\rangle$ is the thermally averaged mean-square displacement for atom i in the protein; the latter quantity is proportional to the crystallographically determined Debye-Waller factor if static disorder is neglected (see Chapt. VI). To simplify the treatment, average mean-square displacements can be used to represent the different types of atoms. The factor $S(\mathbf{r}_i)$ is an empirical scaling function that accounts for the interatomic screening of particles which are away from the RZ-RR boundary;[108] it varies from 0.5 at the reaction zone boundary to zero at the reaction region (see Fig. 8).

The dynamics simulation is limited to the atoms in the reaction zone. Atoms in the reaction region are treated by ordinary molecular dynamics and their motions are governed by Newton's equations of motion. Atoms in the buffer region, as indicated above, obey a Langevin equation of motion. Thus we have a set of simultaneous equations

$$m_i\ddot{\mathbf{r}}_i(t) = \begin{bmatrix} -\nabla_i[U(\mathbf{r}_1, \mathbf{r}_2, \ldots, \mathbf{r}_N)] + \mathbf{F}_B(\mathbf{r}_i) & \text{(reaction region)} \\ -\nabla_i[U(\mathbf{r}_1, \mathbf{r}_2, \ldots, \mathbf{r}_N)] + \mathbf{F}_B(\mathbf{r}_i) \\ \qquad - \beta_i m_i\mathbf{v}_i(t) + m_i\mathbf{f}_i(t) & \text{(buffer region)} \end{bmatrix} \quad (14)$$

where definitions for $\mathbf{F}_B$(solvent) and $\mathbf{F}_B$(protein) are given by Eqs. 12 and 13, respectively; the Langevin forces, $m_i\mathbf{f}_i(t)$, are defined in Eqs. 10 and 11, and the coordinates $(\mathbf{r}_1, \mathbf{r}_2, \ldots, \mathbf{r}_N)$ refer to the N atoms in the reaction zone.

These modified equations of motion are solved numerically using a Verlet type of algorithm which includes the effect due to the presence of Langevin forces[99,102] (see Chapt. IV.G).

There are some limitations to the stochastic boundary molecular dynamics approach in its present form. Since the method is limited to a local region, it neglects explicit effects of the rest of the system. Extensions of the theory to include the influence of low-frequency motions and fluctuating long-range electrostatic interactions on the local dynamics are possible. Also inherent to reduced dynamics descriptions is the introduction of information regarding the characteristics of the neglected part of the system. This information is contained in the solvent-solvent distribution functions, structural and thermal fluctuation parameters for the protein atoms, and the Langevin parameters (friction coefficients) for the buffer region atom. Thus preliminary simulations or empirical rules must be available to obtain these results prior to application of the method to the system of interest.

D. STOCHASTIC DYNAMICS WITH A POTENTIAL OF MEAN FORCE

In many instances one is interested only in the motion of a portion of the macromolecule, or of a peptide system, but wishes to include the effect of solvent on these motions. This is the case for a study of the stability and dynamics of a protein in a range of solvent environments. Alternatively, one may wish to represent one part of a protein (e.g., a sidechain) as moving in a "solvent" bath provided by the remainder of the protein. Techniques that reduce the magnitude of the problem by eliminating the explicit solvent degrees of freedom from the calculation are appropriate for such studies. The simulation problem is thereby reduced essentially to the labor of a vacuum simulation. To do this in a meaningful way, both the equilibrium and the dynamic effect of the solvent on the system of interest must be taken into account. The former involves a modification of the potential energy function by the presence of the solvent. This is introduced by use of a potential of mean force, which is defined as the potential whose negative gradient gives the mean force in solution between the particles making up the systems of interest. The potential of mean force $W(\mathbf{r}_1, \mathbf{r}_2, \ldots, \mathbf{r}_N)$ for a system of N particles determines the effective interactions of the N particles fixed at positions $\mathbf{r}_1$, $\mathbf{r}_2, \ldots, \mathbf{r}_N$, where the positions of the solvent particles have been canonically averaged over all configurations. Formally, the potential of mean force is related to the N-particle distribution function $g(\mathbf{r}_1, \mathbf{r}_2, \ldots, \mathbf{r}_N)$; i.e.,

$$g(\mathbf{r}_1, \mathbf{r}_2, \ldots, \mathbf{r}_N) = \exp[-\beta W(\mathbf{r}_1, \mathbf{r}_2, \ldots, \mathbf{r}_N)] \tag{15}$$

The dynamical aspects of the solvent are approximated by including stochastic and dissipative forces in the equations of motion. To treat the effects of solvent in a simple fashion, the Langevin equation

$$m_i \ddot{\mathbf{r}}_i = -\nabla_i[W(\mathbf{r}_1, \mathbf{r}_2, \ldots, \mathbf{r}_N)] - \beta_i m_i \mathbf{v}_i(t) + m_i \mathbf{f}_i(t)$$

$$i = 1, \ldots, N \quad (16)$$

is used with the potential of mean force $W(\mathbf{r}_1, \mathbf{r}_2, \ldots, \mathbf{r}_N)$ and with the dissipative and random forces defined in Eqs. 10 and 11, respectively. In certain applications of the stochastic dynamics approach, the potential of mean force can be obtained by analysis of a full dynamical simulation. For example, the distribution function for one part of a protein (such as the orientation of a sidechain) is evaluated and then used to determine the potential of mean force. Alternatively, when all of the solvent or bath degrees of freedom are to be eliminated, the statistical mechanical theory for the equilibrium structure of polar fluids[109,112-115a] can be employed. In the simplest statistical mechanical approach, a superposition approximation is introduced and the potential of mean force is written in the form

$$W(\mathbf{r}_1, \mathbf{r}_2, \ldots, \mathbf{r}_N) = U(\mathbf{r}_1, \mathbf{r}_2, \ldots, \mathbf{r}_N) + \sum_{i>j} \Delta W(|\mathbf{r}_i - \mathbf{r}_j|) \quad (17)$$

where $U(\mathbf{r}_1, \mathbf{r}_2, \ldots, \mathbf{r}_N)$ is the vacuum potential and ΔW is the solvent contribution to the potential of mean force for the pair of atoms i and j in the molecule of interest;[115] the solvent contribution is directly related (in accord with Eq. 15) to the distribution function $g(|\mathbf{r}_i - \mathbf{r}_j|)$ for the pair i and j at infinite dilution in the solvent. The theoretical approach has been extended to avoid the need for the superposition approximation, which neglects solvent shielding by one part of a molecule on another part, in the calculation of the potential of mean force.[26a,115a]

Stochastic dynamics has been found to be particularly useful for introducing simplified descriptions of the internal motions of complex systems. When applied to small systems (e.g., a peptide or an amino acid sidechain) it is possible to do simulations that extend into the microsecond range, where many important phenomena occur. Simulation studies using this method have been carried out, for example, to explore solvent effects on the dynamics of internal "soft" degrees of freedom in small biopolymers, e.g., the dynamics of dihedral angle rotations in the alanine dipeptide (see Chapt. IX.B.1).

Clearly, the present approach precludes the detailed study of the dynamics of explicit solute-solvent interactions because the solvent (bath) degrees of freedom have been eliminated. Also, as in the stochastic boundary model, one is required to introduce information about the system beyond the potential energy functions required for a full molecular dynamics simulation; i.e., the stochastic parameters and the calculation of the potential of mean force are required. In addition, due to the use of the simple Langevin description for the dynamical heat bath effects, the method is limited to the study of larger-scale, low-frequency motions, i.e., the motions must be slow relative to the relaxation times of the solvent. However, extensions to faster processes, with, for example, frequency-dependent friction coefficients, are possible.[116]

E. ACTIVATED DYNAMICS

The time scale of many chemical and physical processes occurring in biomolecules is limited by the rate of overcoming an energy barrier. One example occurs in the binding of oxygen to myoglobin, where the ligand must pass several energy barriers of varying size before arriving at the binding site.[117] Another is provided by the well-studied case of the 180° rotation of aromatic sidechains ("ring flips") in proteins (see Chapt. VI.B.1).[118-120] Most enzyme-catalyzed reactions also involve barrier crossing; triosphosphate isomerase is one such case for which detailed experimental data are available.[121] The phenomenological time scale of such activated events is often as long as a microsecond; i.e. while these processes can be intrinsically fast, they occur only infrequently (with an average frequency of 10^{11} s^{-1} or less). Thus they are not adequately sampled in conventional simulation approaches. Activated dynamics methods provide one alternative that overcomes this sampling problem, although stochastic dynamics has also been applied to barrier crossing phenomena (see Chapt. IX.B.3).

Special simulation techniques have been developed which make possible the efficient determination of trajectories of relevance in calculating the rates for activated processes.[122-124] These techniques, which were earlier applied to small-molecule collision dynamics[125-127] and to vacancy diffusion dynamics in regular solids,[128] represent a synthesis of molecular dynamics methods and transition-state theory that can be used with the more general dynamics techniques discussed above. In such calculations, a "reaction coordinate," ξ, typically a set of n atomic coordinates which carry the system from a reactant configuration to a product configuration, is first identified and the free-energy change associated with adiabatically stepping along the reaction coordinate by the minimum free-energy path connecting reactant and product states is determined. This calculation may be carried out using approximate adiabatic mapping techniques, special methods for identifying reaction paths[128a] or umbrella sampling procedures.[129] In the adiabatic mapping approach, one calculates the minimum energy of the protein consistent with a given structural change.[130] Minimization allows the remainder of the protein to relax in response to the structural change, so that the resulting energy provides an approximation to the potential of mean force for the reaction path, $W(\xi)$. Accurate potentials of mean force can be calculated using the umbrella sampling methods. To do this, a series of constraining ("window") potentials, $V_i(\xi)$, are constructed to bias conformations toward values of ξ in the neighborhood of a set of values ξ_i. The biased probability density, $\rho_i^*(\xi)$, is then computed for each $V_i(\xi)$ by a simulation with no restrictions on the other coordinates of the system. The actual probability density $\rho(\xi)$ is related to the $\rho_i^*(\xi)$ by the expression

$$\rho(\xi) = e^{\beta V_i(\xi)}\langle e^{-\beta V_i(\xi)}\rangle \rho_i^*(\xi)$$

where $\langle e^{-\beta V_i(\xi)}\rangle$ is the average of the umbrella potential for window i over the unbiased distribution function. Since $\langle e^{-\beta V_i(\xi)}\rangle$ is not determined directly by the simulation, a method based on fitting together the $\rho_i^*(\xi)$ from a series of overlapping windows is used to obtain relative values of $\rho(\xi)$. To make this explicit, it is helpful to introduce $W(\xi)$, defined by

$$W(\xi) = -k_B T \ln \rho(\xi)$$

where $W(\xi)$ is the free energy as a function of the reaction coordinates, the potential of mean force for the coordinate ξ (see Chapt. IV.D, above). The probability density $\rho_i^*(\xi)$ is related to $W_i(\xi)$ by

$$W_i(\xi) = -k_B T \ln \rho_i^*(\xi) - V_i(\xi) + C_i$$

where $C_i = -k_B T \ln \langle e^{-\beta V_i(\xi)}\rangle$. With $\rho_i^*(\xi)$ evaluated by the simulations and $V_i(\xi)$ a known function, $W_i(\xi) + C_i$ can be calculated for the overlapping window and the results fitted together to obtain a continuous function that approximates $W(\xi)$, from which $\rho(\xi)$ can be obtained. The point of highest free energy along this reaction path, $\xi\dagger$, is the transition state and $\rho(\xi\dagger)/\int_i \rho(\xi)\,d\xi$ is the probability of "finding" the system at the top of the barrier, where the integral is over values of the reaction coordinate corresponding to the initial-state valley.[124]

With a knowledge of $\rho(\xi)$, the rate constant k for the reaction can be written[124]

$$k = \tfrac{1}{2}\kappa \langle |\dot{\xi}|\rangle_{\xi\dagger}\, \rho(\xi\dagger)\Big/ \int_i \rho(\xi)\,d\xi \tag{18}$$

The quantity $\langle |\dot{\xi}|\rangle_{\xi\dagger}$ is the average absolute value of the crossing velocity, $\dot{\xi} \equiv d\xi/dt$, evaluated at $\xi\dagger$, and κ is the transmission coefficient.

If κ is set equal to unity and the equilibrium value is used for $\langle|\dot{\xi}|\rangle_{\xi\dagger}$, the rate constant reduces to that obtained from transition-state theory.[127,131] However, deviations from the ideal transition-state rate often occur so that the reactive flux is, in general, not equal to this simple result, which is determined by the equilibrium properties of the system. It is then necessary to evaluate the transmission coefficient, which accounts for the probability of multiple crossings, and to calculate the (nonequilibrium) velocity distributions at the transition state. Both of these quantities may be computed from trajectories that originate at the transition state. An ensemble of such transition-state configurations (on the order of 500 may be needed for adequate statistics) is constructed by constraining the reaction coordinate to the neighborhood of its transition-state value, $\xi\dagger$, (e.g., by use of an umbrella sampling method), while allowing the remaining coordinates to evolve in accord with the appropriate distribution function (e.g., by solving the equations of motion for the constrained system). Trajectories are then calculated for each member of the ensemble without constraining the reaction coordinate, and the dynamics of the reaction coordinate are followed from the transition state both forward and backward in time. From the resulting ensemble of trajectories the average absolute value of the velocity of the reaction coordinate $\langle|\dot{\xi}|\rangle_{\xi\dagger}$ and the reactive flux correlation function, $\kappa(t)$, can be computed. The quantity $\kappa(t)$ is obtained from the expression[131]

$$\kappa(t) = D\langle\dot{\xi}(0)\delta[\xi(0) - \xi\dagger]H[\xi(t)]\rangle \tag{19}$$

Here D is a normalization constant which ensures that $\kappa(0_+) = 1$, $\delta(x)$ is a Dirac delta function, and H is a step function that is equal to 1 for $\xi > \xi\dagger$ and zero otherwise. In most cases $\kappa(t)$ approaches a plateau value within a very short time (within a ps or less) that can be identified with the transmission coefficient κ in Eq. (18).[131a]

The major difficulty in applying the activated dynamics method is the determination of an optimal reaction coordinate and the transition state for complicated many-body systems, such as a reacting enzyme-substrate complex in solution. This problem is rooted in the conceptual and computational complexity associated with finding the minimal number of atomic coordinates which adequately specify the transition from a "reactant" configuration to a "product" configuration.[124,131,132] A poor choice of reaction coordinate or transition state does not necessarily invalidate the method. However, the efficiency of the activated dynamics–transition state sampling rapidly decreases as the reaction coordinate become less than optimal. It is here that methods for reducing the effective size of the system (e.g., molecular dynamics with stochastic boundaries or stochastic dynamics with a potential of mean force) and improvements in the reaction flux methodology,[132a] can help in making tractable calculations for complicated biochemical processes. Another important problem in modelling reactions is the determination of the potential energy surface, particularly for complex systems; the development of combined classical and quantum mechanical potential functions should be helpful in this regard.[132b]

F. HARMONIC AND QUASI-HARMONIC DYNAMICS

Normal coordinate analysis has been used for many years in the interpretation of vibrational spectra for small molecules.[88] It provided the motivation for the application of the harmonic approximation to proteins and their constituent elements (e.g., an α-helix).[36,133-136] In this alternative to conventional dynamical methods, it is assumed that the displacement of an atom from its equilibrium position is small and that the potential energy (as obtained from Eq. 6) in the vicinity of the equilibrium position can be approximated as a sum of terms that are quadratic in the atomic displacements; i.e., making use of Cartesian coordinates, which are simplest to employ for large molecules, we have

$$U(\mathbf{r}_1, \mathbf{r}_2, \ldots, \mathbf{r}_N) \doteq \tfrac{1}{2} \sum_m \sum_n (\mathbf{r}_m - \mathbf{r}_m^{eq})$$

$$\cdot \left. \frac{\partial^2 U(\mathbf{r}_1, \mathbf{r}_2, \ldots, \mathbf{r}_N)}{\partial \mathbf{r}_m \, \partial \mathbf{r}_n} \right|_{\mathbf{r}_i = \mathbf{r}_i^{eq}(i \neq m,n)}$$

$$\cdot (\mathbf{r}_n - \mathbf{r}_n^{eq}) + O[(\Delta \mathbf{r})^3]$$

$$= \tfrac{1}{2} \sum_m \sum_n (\mathbf{r}_m - \mathbf{r}_m^{eq}) \cdot \mathbf{K}_{mn} \cdot (\mathbf{r}_n - \mathbf{r}_n^{eq}) \qquad (20)$$

where $(\mathbf{r}_m - \mathbf{r}_m^{eq})$ corresponds to the displacement of atom m from its equilibrium position and the K_{mn} are the elements of the force constant matrix, $\mathbf{K}$, given by the second derivatives of the potential energy evaluated at the equilibrium (minimum energy) geometry. For a protein, the structure used is usually obtained by minimization with an empirical energy function starting with the known X-ray structure; a local minimum is utilized since global minimization of a function of many variables is very difficult (see Chapt. IV.H). The force constants together with the atomic masses can be used to set up a $3N$ by $3N$ matrix for determining the normal vibrational modes of the molecule.[88] Solution of the resulting matrix equations by diagonalization of the mass-weighted force constant matrix yields a set of normal frequencies ω_i ($i = 1$, $3N$) and the $3N$ associated normal-mode eigenvectors. Six of these modes are associated with eigenvalues of zero frequency and correspond to the translations and rotations of the entire molecule. The remaining $3N - 6$ modes and frequencies provide details of the internal dynamics of the N-atom system within the quadratic (harmonic) approximation.

The time development of the displacements in the harmonic approximation for atom n of a molecule in thermodynamic equilibrium at a temperature T has the analytic form

$$\Delta \mathbf{r}_n(t) = \sum_{i=7}^{3N} \left(\frac{k_B T}{m_n}\right)^{1/2} \frac{\alpha_n^i}{\omega_i} \cos(\omega_i t + \phi_i) \tag{21}$$

where a (random) phase shift ϕ_i is introduced for each mode; m_n is the mass of atom n and α_n^i is the vector of the projections of the ith normal mode, with frequency ω_i, on the Cartesian components of the displacement vector for the nth atom. From the $\Delta \mathbf{r}_n(t)$, which provide the trajectory of the harmonic system, the equilibrium and dynamic properties can be computed for all times. For example, the mean-square displacements of the atoms from their equilibrium position are given by

$$\langle (\Delta \mathbf{r}_n)^2 \rangle = \sum_{i=7}^{3N} \frac{k_B T}{m_n} \left| \frac{\alpha_n^i}{\omega_i} \right|^2 \tag{22}$$

and the time-dependent displacement correlation functions are of the form

$$\langle \Delta \mathbf{r}_n(t) \Delta \mathbf{r}_n(0) \rangle = \sum_{i=7}^{3N} \frac{k_B T}{m_n} \left| \frac{\alpha_n^i}{\omega_i} \right|^2 \cos(\omega_i t) \tag{23}$$

Although the harmonic model does not provide a complete description for the motional properties when anharmonic contributions are important, it is a useful first approximation because of its simple analytic form. Normal mode analyses can be compared directly with vibrational spectra obtained from infra-red, Raman or inelastic neutron scattering data. Further, the harmonic model is of considerable importance for the calculation of the motional contribution to thermodynamic properties, such as the heat capacity, absolute entropy, and free energy (see Chapt. V). The harmonic approach provides the most direct method of computing these thermodynamic properties for systems in which quantum corrections are essential. This is the case for most molecules, including proteins, at ordinary temperatures. In addition, harmonic dynamics has been found to be a useful tool in exploring the motions of α-helices,[133,134,137] β-sheets, and a number of small proteins,[135,136a] as well as nucleic acid oligomers.[138]

An alternative to harmonic dynamics which incorporates some effects due to the anharmonic nature of the forces is called quasi-harmonic dynamics and is related to approaches used earlier in solid-state physics.[139] The quasi-harmonic model for macromolecules was first suggested for evaluating the temperature-dependent anharmonic corrections to the internal entropy.[140] In the quasi-harmonic approach the force constant matrix is constructed from the second moments of the atomic displacements obtained from a molecular dynamics or Monte Carlo simulation. The second moments of the displacements of atoms m and n, σ_{mn}, have the Cartesian components $[\sigma_{mn}]^{\alpha\beta}$,

$$[\sigma_{mn}]^{\alpha\beta} = \langle (\mathbf{r}_m - \langle \mathbf{r}_m \rangle)^\alpha (\mathbf{r}_n - \langle \mathbf{r}_n \rangle)^\beta \rangle \tag{24}$$

where the angular brackets represent averages over the simulation. The quasi-harmonic force constant matrix, $\mathbf{K}^Q$, can then be written

$$\mathbf{K}^Q = k_B T \sigma^{-1} \tag{25}$$

where σ is the second moment matrix for the entire molecule. The quasi-harmonic eigenvalues and eigenvectors are calculated from the force constant matrix in the same way as in the harmonic approximation. The resulting quasi-harmonic modes and frequencies may then be used to compute the full range of dynamic and thermodynamic properties of the system. The quasi-harmonic approximation permits one to find a temperature-dependent correction to the harmonic limit, and its utility for the calculation and interpretation of such properties as internal entropy and free energy[140-142] has been demonstrated. However, its validity is limited to cases where the anharmonic correction is small; for motions in double minimum potentials, for example, the quasi-harmonic approximation is not appropriate. Here path integral[143] and related methods for determining quantum corrections in anharmonic systems may be of use, although they have not yet been applied to proteins.

G.   ALGORITHMS FOR MOLECULAR AND STOCHASTIC DYNAMICS

A variety of algorithms have been used for integrating the equations of motion in molecular dynamics simulations of macromolecules. Most widely employed are the algorithms due to Gear[91] and Verlet.[90] The algorithm introduced by Verlet in his initial studies of the dynamics of Lennard-Jones fluids is derived from the two Taylor expansions,

$$\mathbf{r}_i(t \pm \Delta t) = \mathbf{r}_i(t) \pm \mathbf{v}_i(t)\,\Delta t + \frac{\mathbf{F}_i(t)}{m_i}\frac{(\Delta t)^2}{2!} \pm \frac{\dot{\mathbf{F}}_i(t)}{m_i}\frac{(\Delta t)^3}{3!} + O[(\Delta t)^4] \tag{26}$$

Their sum yields the algorithm for propagation of the positions,

$$\mathbf{r}_i(t + \Delta t) = 2\mathbf{r}_i(t) - \mathbf{r}_i(t - \Delta t) + \frac{\mathbf{F}_i(t)}{m_i} (\Delta t)^2 + O[(\Delta t)^4] \quad (27)$$

and their difference yields the algorithm for propagation of the velocities,

$$\mathbf{v}_i(t) = \frac{\mathbf{r}_i(t + \Delta t) - \mathbf{r}_i(t - \Delta t)}{2 \Delta t} + O[(\Delta t)^3] \quad (28)$$

In the equations above, Δt represents the time step, $\mathbf{F}_i(t)$ is the force on atom i at time t, and m_i is the mass of atom i. The algorithm embodied in Eqs. 27 and 28 provides a stable numerical method for solving Newton's equations of motion for systems ranging in complexity from simple fluids to biopolymers. One should note that in this algorithm the velocities play no role in propagating the position at time t to that at time $t + \Delta t$. Thus the Verlet algorithm must be modified in order to incorporate velocity-dependent forces or temperature scaling. Also, the Verlet algorithm is not "self-starting," and a lower-order Taylor expansion [terms to $O\{(\Delta t)^2\}$] is often used in initiating the calculation. Modified forms of the Verlet algorithm are the "leapfrog" algorithm,[144] and the Beeman algorithm,[145] which are obtained by similar manipulations; the former appears to have greater stability in some applications. Both the Beeman algorithm and the leapfrog algorithm yield a position propagation scheme that is identical to the Verlet algorithm; they differ in that the velocity appears explicitly in the propagation of the position. The Gear algorithm[91] includes higher-order corrections [$O\{(\Delta t)^5\}$] than does the Verlet algorithm, but for systems with many internal degrees of freedom it has been found that the increase in calculation required for each step is not compensated by the increase in step size. For protein calculations, a step size that yields valid results is on the order of 1 fs, although somewhat larger step sizes (2 to 4 fs) have been reported.[23] The step size of 1 fs is appropriate, as well, for treating the aqueous solvent explicitly if the water molecules are constrained to have a fixed geometry.[146]

For the addition of velocity-dependent forces, such as the dissipative, Langevin force in Eq. 10, an algorithm may be derived which reduces directly to the Verlet algorithm in the limit of vanishing friction ($\beta_i \rightarrow 0$). This algorithm is obtained by adding the Langevin terms to Eq. 27 and substituting with Eq. 28 for $\mathbf{v}_i(t)$. An algorithm of order $(\Delta t)^3$ results that is valid for $\beta_i \Delta t < 1$. The propagation equation in one dimension has the form

$$x_i(t + \Delta t) = \left\{2x_i(t) - x_i(t - \Delta t) + \left[\frac{F_i(t)}{m_i} + f_i(t)\right](\Delta t)^2\right.$$

$$\left. + \tfrac{1}{2}\beta_i \Delta t x_i(t - \Delta t)\right\}(1 - \tfrac{1}{2}\beta_i \Delta t)^{-1} \qquad (29)$$

In this equation x_i is the Cartesian coordinate x for atom i, β_i is the frictional drag on atom i and f_i is the Langevin random force on atom i obtained from a Gaussian random distribution of zero mean and variance

$$\langle f_i f_j \rangle = \frac{2k_B T}{m_i} \frac{\beta_i}{\Delta t} \delta_{ij} \qquad (30)$$

This algorithm has been used in the integration of the Langevin equation applied to the buffer zone atoms in the stochastic boundary molecular dynamics method (Chapt. IV.C), as well as in other stochastic dynamics calculations.[102]

For some molecular dynamics simulations (e.g., in the constant-temperature approaches discussed in Chapt. IV.B) the algorithm for the atomic displacement depends explicitly on the velocity and methods different from the Verlet algorithms are required; possible choices are the leapfrog and Beeman algorithms (see above). The algorithms used can usually be expressed in a form that is optimal for the particular dynamics. Examples of the introduction of such alternative algorithms may be found in work on molecular dynamics simulations of NPT and NVT ensembles[96] and on the numerical integration of general stochastic equations of motion.[147] In addition to algorithms that propagate the equations of motion with thermodynamic constraints on the system, methods have been introduced to hold bond lengths, and sometimes bond angles, fixed at their ideal values. The most commonly used of these methods is the SHAKE algorithm.[146] When applied to constrain bond distances in a protein minimization or dynamics simulation, the SHAKE algorithm provides some savings in computer time because a larger step size can be used.[146,148]

For some problems, such as the motion of heavy particles in aqueous solvent (e.g., conformational transitions of exposed amino acid sidechains, the diffusional encounter of an enzyme-substrate pair), either inertial effects are unimportant or specific details of the dynamics are not of interest; e.g., the solvent damping is so large that inertial memory is lost in a very short time. The relevant approximate equation of motion that is applicable to these cases is called the Brownian equation of motion,

$$\mathbf{v}_i(t) = \frac{1}{k_B T} \sum_j \mathbf{D}_{ij} \mathbf{F}_j(t) + \mathbf{R}_i(t) \qquad (31)$$

with

$$\langle \mathbf{R}_i(t) \rangle = 0 \quad \text{and} \quad \langle \mathbf{R}_i(t)\mathbf{R}_j(0) \rangle = 2\mathbf{D}_{ij}\delta(t) \tag{32}$$

In these equations $\mathbf{D}$ represents the general diffusion tensor for interacting particles, which may include hydrodynamic interactions. To simulate such Brownian motion, an efficient algorithm based on Eqs. 31 and 32 has been developed.[149] This algorithm, written in its most general form, which accounts for interparticle (hydrodynamic) velocity-dependent interactions as well as direct interactions, is

$$\mathbf{r}_i(t + \Delta t) = \mathbf{r}_i(t) + \mathbf{R}_i(t) + \frac{\Delta t}{k_B T} \sum_{j=1}^{N} \mathbf{D}_{ij}[(r_1(t), \ldots, r_N(t)]\mathbf{F}_j(t) \tag{33}$$

where the possible dependence of the diffusion tensor on the N-particle configuration is displayed explicitly. In correspondence with Eq. 32, the random displacement, $\mathbf{R}_i(t)$, is sampled from a zero-mean, Gaussian distribution with variance

$$\langle \mathbf{R}_i \mathbf{R}_j \rangle = 2\mathbf{D}_{ij} \Delta t \tag{34}$$

Brownian dynamics algorithms have been used to explore a number of slow processes in systems containing biopolymers. They include numerical simulations of local folding and unfolding,[150,151] large-amplitude fluctuations in multilobed proteins,[152,153] and the calculation of rate constants for the association of biopolymers; these applications are described in Chapts. VII–IX.[154,155]

H. MINIMIZATION ALGORITHMS

A technique of very general use in the study of macromolecules of biological interest, as well as of smaller inorganic and organic molecules, is the method of coordinate (or geometry) optimization often called energy minimization. Although energy minimization is not a dynamical method, per se, it plays such an important role (e.g., in starting dynamics simulations) that a brief description is necessary to complete the methodological discussion. There is a vast mathematical literature on locating multidimensional extrema of a function of many variables.[156] In this section we outline some of the methods relevant to globular proteins. The essential problem is that of finding the coordinates that minimize the potential energy (as given by a function such as Eq. 6) of the system of interest. This is an intrinsically difficult problem, particularly for a macromolecule with many degrees of freedom, because the nonlinear nature of the commonly used potential functions leads to numerous minima. Except for small peptides, a full grid search is impossible. Thus alternative approaches, which generally find a local minimum rather than the global minimum, are employed. Since the potential energy function is usually expressed in analytic form, the derivatives are given by simple functions that can be used to expedite the search for a minimum.

For proteins it is common practice to begin with a starting structure that is geometrically close to the desired solution. One starting point is the X-ray structure; another is a configuration obtained from a computer simulation. Given such a set of initial coordinates, energy minimization permits one to find a structure in a neighboring local minimum consistent with the potential function that is employed.

The practical approaches for protein geometry refinement rely on iterative local linearizations of the full nonlinear optimization problem. The procedure is to step along the potential surface in a direction that decreases the energy. Such iterative approaches in Cartesian coordinates may be symbolized by

$$\mathbf{r}_{n+1} = \mathbf{r}_n + \delta_n \tag{35}$$

where the subscripts refer to the number of the iterate, and δ_n is the nth displacement in the configuration space of the system. We first discuss the simpler methods, which employ only first-derivative information to obtain δ_n, and then present techniques that also use the second derivatives.

The two most commonly used first-derivative approaches are the method of steepest descent[157] and the conjugate gradient method.[158] For both of these, the entire vector of first partial derivatives of the potential energy with respect to all $3N$ coordinates is needed.

In the steepest-descent method, a displacement opposite to the potential energy gradient (i.e., in the direction of the force) is added to the coordinates at each step. This can be written

$$\delta_n = -k_n(\nabla_n V_T) \tag{36}$$

where V_T is the total potential energy (Eq. 6) and k_n is a parameter that adjusts the step size to take account of the fact that the energy may increase, as well as decrease, after a step is taken (e.g., if the energy decreases, k_n is increased for the next step, while if the energy increases, presumably because the step size was too large, k_n is decreased). Although the steepest-descent method suffers from poor convergence, it has the important property that the minimized structure has the smallest displacement from the starting configuration. This is true because the method goes directly to the nearest minimum.

Although such an approach rapidly relieves bad van der Waals contacts and strained bond lengths and bond angles, it does not locate nearby minima that may be substantially lower in energy but are separated from the steepest descent minimum by a barrier. The steepest-descent method is good for relieving strain in a starting geometry, but it is not an efficient method for finding the minimum, particularly on the complex potential energy surfaces that characterize most macromolecules.

A more sophisticated approach, which also uses only first-derivative information, is the conjugate gradient method.[148,158] It has considerably better convergence characteristics than the steepest-descent method. The conjugate gradient algorithm makes use of the previous history of minimization steps as well as the current gradient to determine the next step. In addition, the step size δ_n is modulated by a parameter, α, which is chosen to give the optimal step; a frequently used method to determine α is a simple line search, which requires a few extra energy evaluations per step.

Symbolically, the conjugate gradient algorithm can be written in terms of the parameters δ_n and α in the form

$$\delta_n = -\mathbf{g}_n + \delta_{n-1} \frac{|\mathbf{g}_n|^2}{|\mathbf{g}_{n-1}|^2} \qquad \mathbf{r}_{n+1} = \mathbf{r}_n + \alpha\delta_n \qquad (37)$$

where $\mathbf{g}_n = \mathbf{\nabla}_n V_T$ denotes the gradient vector for the nth coordinate set. Equation 37 shows that the conjugate gradient technique, as do related methods, such as the Powell algorithm,[159] makes a given step a linear combination of the current gradient and the previous step. For a N-dimensional quadratic surface the conjugate gradient method reaches the minimum in, on the order of, N steps. When the algorithm is found to be making little progress on a nonquadratic surface, it may be reinitialized by setting the contribution from previous steps ($\mathbf{g}_n$ in Eq. 37) to zero and continuing. Although this method does require more energy evaluations per step than the steepest-descent algorithm, it usually converges more rapidly and often produces a substantially lower energy when it has converged.

Introduction of second derivative information in the energy minimization procedure improves the rate of convergence in many cases. In particular, if the potential energy surface has a quadratic dependence on the displacement from the minimum, it is possible to start at any point $\mathbf{r}_0$ and arrive at the extremum, $\mathbf{r}_{min}$, in one step; i.e., in one dimension for simplicity, we have

$$x_{min} = x_0 - \left[\frac{dV_T}{dx}\right]_{x_0} / \left[\frac{d^2V_T}{dx^2}\right]_{x_0} \qquad (38)$$

Use of Eq. 38 on a nonquadratic surface in an iterative fashion forms the basis of the Newton-Raphson algorithm.[160] For the multidimensional case, the matrix of second derivatives is called the Hessian matrix $\mathbf{H}$. For the nth coordinate set it is defined as

$$[\mathbf{H}_n]_{k,l} = \frac{\partial^2 V_T}{\partial x_k\, \partial x_l}\bigg|_n \tag{39}$$

Making use of Eq. 38, we find that δ_n in Eq. 35 can be written

$$\delta_n = -\mathbf{H}_n^{-1}\mathbf{g}_n \tag{40}$$

Near a minimum, where the potential is expected to be approximately quadratic, the Newton-Raphson algorithm leads to rapid convergence. Far from a minimum, it may be inefficient and in some cases even pathological in behavior when the surface is far from quadratic (e.g., it may increase the energy). In addition, this method requires extensive amounts of computer memory and time for large systems, due to the requirement for construction and inversion of the Hessian matrix. In some cases it may be possible to simplify the calculation by keeping only the largest elements of the Hessian matrix.[160a]

A more generally useful second-derivative method that is particularly suited for large systems such as proteins is the adopted basis Newton-Raphson (ABNR) algorithm.[65] Rather than using the full multidimensional set of basis vectors, a basis is adopted that is limited to the subspace in which the system has made the most progress in the past moves. At the nth step of the iteration for a subspace of dimension p the basis vectors are comprised of the difference of the current vector with the last p position vectors; that is, the basis vectors are taken to have the form

$$\begin{aligned} \Delta\mathbf{r}_n^1 &= \mathbf{r}_{n-1} - \mathbf{r}_{n-p-1} \\ &\vdots \qquad\quad \vdots \\ \Delta\mathbf{r}_n^p &= \mathbf{r}_{n-1} - \mathbf{r}_{n-2} \end{aligned} \tag{41}$$

Usually, p is chosen to be a number between 4 and 10. In this way the system moves in the best direction in a restricted subspace. For this subspace the second-derivative matrix is constructed by finite differences from the stored displacement and first-derivative vectors and the new positions are determined as in the Newton-Raphson method. This method is quite efficient in terms of the required computer time, and the matrix inversion is a very small fraction of the entire calculation. The adopted basis Newton-Raphson method is a combination of the best aspects of the first derivative methods, in terms of speed and storage requirements, and the more costly full Newton-Raphson technique, in terms of introducing the most important second-derivative information. Since the method is not self-starting, the first steps are taken with a first-derivative method and steepest descent is the most obvious choice.

A comparison of the results obtained with the various minimization algorithms described in this section, as well as with quenched dynamics, is given in Ref. 65 where applications to a small peptide and a protein are presented.

CHAPTER V

THERMODYNAMIC METHODS

Dynamical techniques are useful for evaluating thermodynamic properties as well as for understanding the internal motions of complex systems. In this chapter we describe methods that can be applied to thermodynamic problems, such as the stability of proteins and their interactions with ligands and substrates. Of essential importance is the calculation of the free energy or entropy of the system, as well as the energy and enthalpy. In most cases the focus is on the difference in thermodynamic properties for two states; examples include two protein conformations (e.g., native versus denatured), wild-type versus mutant proteins, and the protein plus a ligand in solution versus the bound protein-ligand system. These differences can often be treated in detail, even when it is not possible to calculate the individual quantities. In what follows, we often use ligand binding as an example to facilitate the description of the methodology. Applications to other problems can be made in a corresponding fashion.

Because of the time-consuming nature of full dynamical (or Monte Carlo) treatments of the thermodynamic properties of molecules as large as proteins, simplified approaches have been used to obtain approximate results. In what follows, a series of methods is described in increasing order of sophistication. The first section treats classical vacuum calculations, which are concerned with the evaluation of the system energy. Next, methods that take into account internal flexibility and harmonic fluctuations are outlined. Finally, techniques for calculating the free energies in condensed phases are presented.

A. VACUUM CALCULATIONS

The energy of a macromolecule in vacuum can be calculated from the potential energy functions described earlier (Chapt. III). If two conformations of a peptide or a protein are of interest, the energy difference between them can be determined directly. Correspondingly, for enzyme-substrate or receptor-hormone interactions, the energies of the separated and the liganded systems can be compared. To carry out such calculations it is necessary that the structures under consideration be known. In some cases, the information is available

59

from X-ray crystallography; e.g., certain enzyme structures have been determined for the isolated protein and the protein-inhibitor complex. When the structures are not known, they have to be obtained by use of modeling procedures.

If we focus on the example of ligand binding, the first step in determining the interaction energy is to bring the two molecules together. If both molecules are treated as rigid systems, a combination of molecular graphics and "docking" algorithms, including energy minimization or other simulation techniques (e.g., quenched dynamics), can be used to search for the relative geometry that leads to a stable structure. It is in this way that some of the earliest studies of substrate and inhibitor binding were made. An improved interaction energy can be obtained by treating the substrate as flexible and minimizing the energy of the complex, including not only the relative positions of the enzyme-substrate pair but also the internal degrees of freedom of the substrate. Here, the reference state for comparison with the complex is slightly more complicated than in the rigid model, where only the interaction energy has to be considered. If the internal geometry of the substrate is allowed to change on binding, a separate energy minimization must be done for the isolated substrate molecule. In this case the binding energy is

$$\Delta E_B = E_{pl} + (E_l^b - E_l^f) \tag{42}$$

where E_{pl} is the interaction energy of the minimized protein-ligand (substrate) complex, and E_l^b and E_l^f are the internal energies of the bound and free minimized ligand molecule, respectively. It is important to note that only the interaction energy and energy differences for the ligand appear in the calculation. Correspondingly, in the analysis of two conformations of a given molecule only the energy difference between the two would be required. This is an essential point since absolute energy values obtained from an empirical potential energy function are arbitrary; e.g., E_l^f is not related in any way to a measurable quantity but is expressed in terms of an arbitrary zero of energy. The quantity E_l^f is generally smaller than E_l^b (i.e., the positive internal energy of the bound ligand is larger than that of the free ligand), so that the binding energy is smaller in magnitude than E_{pl}. An early example of this type of calculation is the study of the interaction of lysozyme with its substrate.[160b]

A more sophisticated vacuum treatment is achieved by introducing flexibility into the receptor or enzyme as well as the ligand. This is important even if crystal structures are available because the potential functions are not exact, so that the most meaningful results are obtained if the components are being compared at their minimum-energy positions. For the enzyme-ligand system, this requires minimizing the structures of the complex and of the separated species. The resulting interaction energy is given by

$$\Delta E_B = E_{pl} + (E_l^b + E_l^f) + (E_p^b - E_p^f) \qquad (43)$$

where E_p^b and E_p^f are the analogue for the enzyme or receptor of E_l^b and E_l^f. Many static calculations of this type have been made and some useful results have been obtained.[160c]

Although the energy calculations described here are of interest, they have a number of limitations. The first of these is inherent in the inaccuracies of the empirical potential energy functions that are being used. These are known to be significant, as indicated by the sizable difference found between the minimum-energy structure obtained from the potential functions and the observed crystallographic structure, even when the calculations are done for the full crystal system.[161,161a] Such errors can be reduced, in principle, by further refinements of the form of the potential function and the associated parameters.

Further, there are corrections to the thermodynamic properties of a system due to the fact that it is not fixed at the potential energy minimum but is undergoing thermal motion. This means that the average energy is not that associated with the minimum energy structure or even with the average structure, which is the one obtained from an X-ray analysis. Instead, for a molecule or complex that is fluctuating, the average energy corresponds to that which would be obtained by calculating the energy for each one of a series of structures and averaging them with the Boltzmann weights appropriate for the system temperature. This average can be determined by molecular dynamics or Monte Carlo simulations.[162,163]

The free energy of the system also includes entropic contributions arising from the internal fluctuations, which are expected to be different for the separate species and for the liganded complex. These can be estimated from normal-mode analyses by standard techniques,[136,164] or by quasi-harmonic calculations that introduce approximate corrections for anharmonic effects;[140,141] such approaches have been described in Chapt. IV.F. From the vibrational frequencies, the harmonic contribution to the thermodynamic properties can be calculated by using the multimode harmonic oscillator partition function and its derivatives. The expressions for the Helmholtz free energy, A, the energy, E, the heat capacity at constant volume, C_v, and the entropy are (without the zero-point correction)[164]

$$\frac{-A}{Nk_BT} = \sum_i \ln\left[1 - \exp\left(-\frac{\theta_i}{T}\right)\right]^{-1} \qquad (44)$$

$$\frac{E}{Nk_BT} = \sum_i \frac{\theta_i/T}{\exp(\theta_i/T - 1)} \qquad (45)$$

$$\frac{C_v}{Nk_B} = \sum_i \frac{(\theta_i/T)^2 \exp(\theta_i/T)}{[\exp(\theta_i/T) - 1]^2} \tag{46}$$

$$S = -\frac{(A - E)}{T} \tag{47}$$

where N is Avogadro's number, $\theta_i = h\nu_i/k_B$, and ν_i is the frequency of the ith mode. In the limit of low frequency, or equivalently of high temperature, the reduced energy (Eq. 45) and heat capacity (Eq. 46) approach a constant value, whereas the reduced free energy (Eq. 44) and the reduced entropy (Eq. 47) diverge logarithmically. To obtain the vibrational entropy change on ligation, separate calculations for the enzyme, the substrate, and the enzyme-substrate complex would be required. For some comparisons it is necessary to include the zero-point corrections, since the changes in the vibrational frequencies can lead to nonnegligible differences.[29a,136] The harmonic zero-point correction to the reduced energy and free energy is equal to $\Sigma_i \theta_i/2T$.

Other contributions to the thermodynamic properties can also be evaluated in the harmonic oscillator model with the rigid rotor approximation. For any binding process there is the decrease in the entropy of the system that results from the fact that the ligand and protein are brought together into a single complex. Estimates of this entropy change in the simplest (ideal gas) approximation have been made,[165] although the usual assumption of a rigid complex can significantly overestimate the entropy loss.[166] A full vibrational treatment of the complex would take account of the reduction in entropy loss due to the internal flexibility in the vibrational degrees of freedom that correspond to translation or rotation in the separated species.[167]

Vacuum calculations of the thermodynamic properties, including the entropic terms that we have discussed, may yield results that are meaningful if the order of the vacuum free energies for a series of conformations (or substrates in the case of binding) corresponds to that found in solution, even when the individual values are in error. This assumption has often been made but is likely to be valid only in the comparison of two very similar systems (e.g., the binding of two different optical isomers for which the solvation free energies of the substrates in water are identical).[160c] To obtain thermodynamic properties that are meaningful in a quantitative sense, it is essential in most cases to go beyond the vacuum treatment and to introduce the effects of solvation.

B. FREE ENERGIES IN THE CONDENSED PHASE

To introduce the effects of a condensed phase environment, we consider the solvation process for a single molecule as a thermodynamic cycle composed of a series of elementary steps. We first describe the energy involved in the solvation process. The various steps are displayed in the energy diagram given in Fig. 10a. We start in the lower left-hand corner [H_2O(g) + solute(g)] with the solute (e.g., a peptide or protein in a given conformation, a ligand, or a protein-ligand complex) and solvent molecules isolated from each other; each water molecule is also separated from every other water molecule. This state may be taken to be the zero of energy. From this state there are several paths which lead to the total solvation energy of the system. The path vertically upward from the lower left-hand corner yields the total energy of the pure condensed solvent [H_2O(l)] at the temperature and density of interest; this is equal to the energy of condensation of the solvent. The top horizontal path (from left to right) corresponds to forming a cavity in the liquid which is of the correct size to accommodate the solute. This cavity has the solvent in the equilibrium orientation appropriate for accommodating the polar and nonpolar moieties of the solute. The last part of the solvation energy comes from the path on the far right going vertically downward; it represents the energy of interaction between the solute molecule and the properly formed cavity. Since reversible transitions between different thermodynamic states are independent of the path, the arrows can be considered to add like vectors. Thus the total energy associated with the path just described is equal to the horizontal bottom path that goes directly from the gaseous starting materials to the aqueous solvated system. The energy of solution of the solute, or equivalently, the energy of transfer from the gas phase to aqueous solution, can be obtained by following the diagonal path from the upper left to the lower right.

The type of construction given in Fig. 10a, which utilizes the Hess's law of constant heat summation, can serve as a means of quantitatively analyzing the thermodynamics of solvation. Further, this view of the solvation process provides a method for considering different standard states. For nonionic species a commonly used standard state is infinite dilution. Although activities become infinite for ions in this limit, it is still a useful reference state because the analytic Debye-Hückel limiting law is valid in this regime.[168]

To determine the solvation free energy, we have to add the entropic contributions to the energy terms shown in Fig. 10a. Of primary interest in most applications is the difference between the state at the upper left (isolated solute gas and pure liquid solvent) and the state at the bottom right (solute dissolved in the liquid solvent). Considering the upper triangle in Fig. 10a, we find that the cavity formation term in the free energy is exactly zero; that is, the change in the solvent energy resulting from the introduction of the solute (E_{cr} in Fig. 10a) is cancelled exactly by a corresponding term in the entropy ($- TS_{cr}$), at infinite dilution and at finite solute concentrations, as well;[169,169a] i.e., we have

$$E_s = E_{cr} + E_{s'}; \qquad S_s = S_{cr} + S_{s'} \qquad (48)$$

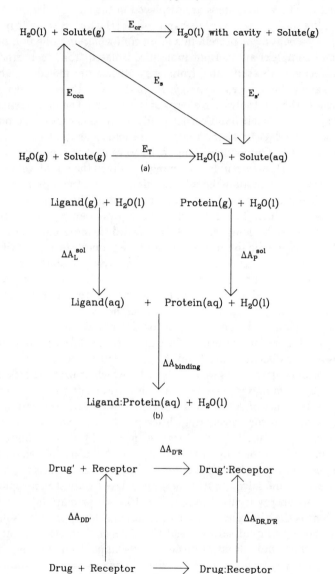

Figure 10. Thermodynamic cycles: (*a*) solvation of a solute; (*b*) solvation and subsequent formation of a protein-ligand complex; (*c*) binding diagram for two different drugs to the same receptor. For a discussion, see the text.

where $E_{s'}$, and $S_{s'}$, are the additional interaction energy and entropy of the solute, and

$$A_s = E_{s'} - TS_{s'} \tag{49}$$

This implies that when the pure liquid solvent is chosen as the reference state, only terms involving canonical averages over the potential energy of interaction between the solute and the solvent (plus the changes in the internal free energies discussed in the previous section) contribute to the free energy of solvation at infinite dilution. At finite concentration, the solute-solute interaction terms have to be considered as well.

In most computer simulations, it is not possible to include enough solvent to approach a system at infinite dilution; it is not uncommon, in fact, to treat solvents at concentrations of tenth molar and higher. Such finite-concentration effects enter into a more complete picture, with the solute modifying the solvent structure and energetics self-consistently. In vivo, one may, in fact, be dealing with a relatively high concentration of molecules other than those of interest, as well as with nonequilibrium systems involving strong concentration gradients. To gain an understanding of the basic aspects of the solvation chemistry, it is useful to examine first systems at equilibrium and to ignore the presence of spectator molecules. As described in Chapt. IV.C, the use of specialized boundary conditions in simulations makes it possible to consider certain types of nonequilibrium effects in solution.

Having shown how the energy and free energy of solution for a single solute can be decomposed, we extend these ideas to ligand binding. Clearly, the effective interactions between the ligand and the protein molecule can be considerably different in aqueous solution as opposed to the gaseous state (i.e., in vacuum). In Fig. 10b, a thermodynamic cycle representing the solvation and subsequent formation of a protein-ligand complex is depicted. Following Fig. 10a, we consider that the ligand and the protein are solvated separately to obtain their equilibrium free energies in solution. If the solutes are flexible, the solvation process can involve modifications in the internal structure and dynamics for both the protein and ligand; this flexibility has been neglected in many studies. In addition to the interactions between the solvated ligand and the protein, the binding involves the displacement of near-neighbor water molecules and ions.

The full calculation of the ligand-protein interaction free energy in solution, as diagrammed in Fig. 10b, corresponds to the determination of the free energy of solvation of the three separate species and the evaluation of the appropriate difference; i.e.,

$$\Delta A_{\text{binding}} = \Delta A_{pl}^{\text{vac}} + \Delta A_{pl}^{\text{sol}} - \Delta A_{l}^{\text{sol}} - \Delta A_{p}^{\text{sol}} + \Delta A_{int}^{\text{sol}} \tag{50}$$

where ΔA_{pl}^{vac} corresponds to the vacuum interaction free energy ΔA_{pl}^{sol}, ΔA_{l}^{sol}, and ΔA_{p}^{sol} represent the free energy of solvation of the protein-ligand complex, the ligand, and the protein, respectively, and ΔA_{int}^{sol} is the difference in the internal free energy between the solvated protein-ligand complex and the separated solvated protein and ligand. The thermodynamics associated with such processes can be determined by statistical mechanics. Computer simulations can provide the desired quantities averaged over the allowed configurations of the system. It is also possible, in some cases, to use more formal approaches such as integral equation methods and to first integrate out the solvent degrees of freedom so as to obtain a solvent-averaged potential or potential of mean force for the ligand-protein interaction. Although, in principle, each of the quantities on the right-hand side of Eq. 50 can be determined by simulation or integral equation methods, in practice the most straightforward calculations involve special cases of the problem depicted in Fig. 10b. For example, instead of determining the full binding free energy, which is generally very difficult for complex systems, it is often possible to focus on the question of the difference in the binding free energy for two similar ligands or two similar proteins (e.g., the wild type and a mutant protein). Such problems involve a thermodynamic cycle corresponding to that shown in Fig. 10c, where the binding of two similar drugs (D and D') to the same receptor is diagrammed. In such a scheme it is not necessary to compute the difference in the free energy of binding from the results obtained for the two horizontal paths. Instead, one uses the unphysical vertical paths, which are ideally suited for thermodynamic perturbation theory. This technique and its implementation by use of simulations or integral equation methods are described below.

C. THERMODYNAMIC PERTURBATION THEORY

The most general approach to the statistical-mechanical problems considered in this chapter is to evaluate the appropriate partition functions for the systems of interest. Given the quantum-mechanical energies, E_k, of the states of the entire system, the partition function has the form

$$Z = \sum_{k} e^{-\beta E_k} \tag{51}$$

with $\beta = 1/k_B T$, where k_B is the Boltzmann constant and T is the absolute temperature. For most simulations of interest for biomolecules the classical analogue of Eq. 51 is adequate,[164] except for some of the stiffer (high-frequency) internal degrees of freedom. Once the partition function has been determined, all macroscopic properties, including free energies, can be derived from it. However, direct evaluation of the partition function for a "macroscopic" system (e.g., a periodic boundary representation, such as is described in Chapt. IV.B) by simulation or other techniques is an intractable problem.[129] Consequently, methods have been devised to calculate the macroscopic properties of interest by evaluating ratios of partition functions.[129] The most widely used approach of this type is the Metropolis Monte Carlo method,[129,170] which like molecular dynamics, is directly applicable to determining accurate energies and enthalpies (e.g., energies of solvation). However, both Monte Carlo and molecular dynamics methods must be extended for evaluating free energies. This is true because the free energy is given by the logarithm of the partition function, $A = -RT \ln Z$, rather than by a derivative of the partition function, as is the energy.

To avoid the need for evaluating the partition function, the focus of free energy calculations is shifted from total free energies to free energy differences. Such calculations are of rather general utility since, as already mentioned, most problems of interest are concerned with differences in the thermodynamic properties of two equilibrium states. Further, any reversible path can be used in going from one state to another. Thus, it is appropriate to determine what is the most convenient path to follow and how one can best compute the change in free energy along that path. Thermodynamic perturbation theory[164] is a very powerful approach to this type of problem. We describe its implementation in terms of simulation and integral equation methods and consider possible reductions of the problem from the many-body case to that of effective interactions involving the potential of mean force between pairs of solutes (e.g., an enzyme and a substrate).

In simulations, an alternative to perturbation theory is umbrella sampling. It is used to connect the two configurations of interest (e.g., a protein plus a bound ligand in solution versus a protein in solution and a free ligand in solution; see Fig. 10(b)) by an appropriate configurational coordinate. Since the calculations correspond exactly to those used with a reaction coordinate in activated dynamics (Chapt. IV.E) we do not repeat the description of the umbrella sampling method.[170a]

The central idea of thermodynamic perturbation theory is that the potential energy function can be partitioned in a convenient way; i.e., one can write

$$V(\mathbf{r}^N; \lambda) = V_0(\mathbf{r}^N) + \lambda V_\lambda(\mathbf{r}^N) \tag{52}$$

where $V_0(\mathbf{r}^N)$ represents the potential for the suitably defined "reference system" and $V_\lambda(\mathbf{r}^N)$ is the perturbation relating the reference system to the system of interest; in Eq. 52, $V_\lambda(\mathbf{r}^N)$ is independent of λ. The symbol $\mathbf{r}^N$ designates the system coordinates and λ is the perturbation parameter, which is unity for the fully perturbated system and zero for the reference state. The linear scaling of the perturbation in the equation above is not necessary but is convenient for the present discussion; more general perturbation-type approaches are also considered in what follows.

The separation in Eq. 52, and the consequences of this separation, described below, are the basis of most calculations of free-energy differences. For one type of calculation, in the spirit of standard perturbation theory, $V_0(\mathbf{r}^N)$ represents a system for which the calculation of the free energy can easily be accomplished (e.g., an uncharged hard-sphere liquid or a volume of noninteracting water molecules) and the perturbed system is the one of interest (a hard-sphere ion in a hard-sphere liquid or a volume of interacting water molecules). A second type of problem is one where the unperturbed system is not necessarily easier to treat than the perturbed system, but it is the difference in free energy between the two that is the quantity of interest. An example would be the drug-receptor problem discussed above, where one drug (or one drug-receptor complex) is the unperturbed system and the other drug (or other drug-receptor complex) is the perturbed system. Alternatively, the problem might concern a single-site mutation in a protein, where one wishes to calculate the free-energy difference between the native protein (unperturbed system) and the mutant protein (perturbed system).

To introduce the formulation, we consider the exact connection between the unperturbed and perturbed systems. We focus on the Helmholtz free energy, A, which is the quantity of interest at constant N, T, and V, where N is the number of particles, T is the temperature, and V is the volume of the system; the alternative case (constant N, T, and P), which leads to the Gibbs free energy, can be treated similarly. The Helmholtz free energy for the potential energy function $V(\mathbf{r}^N; \lambda)$ can be written in terms of the partition function Z_λ as

$$A_\lambda = -(1/\beta) \ln Z_\lambda = C(N, T, V) - (1/\beta) \ln Z_\lambda^c \qquad (53)$$

where $C(N, T, V)$ is a constant related to the kinetic energy portion of the partition function, and the quantity Z_λ^c is the classical configurational partition function,

$$Z_\lambda^c = \int \exp[-\beta V(\mathbf{r}^N; \lambda)] d\mathbf{r}^N \qquad (54)$$

Writing $V(\mathbf{r}^N; \lambda)$ as in Eq. 52, we have

$$Z_\lambda^c = \int \exp[-\beta\{V_0(\mathbf{r}^N) + \lambda V_\lambda(\mathbf{r}^N)\}] d\mathbf{r}^N \qquad (55)$$

which can be expressed in a convenient form by multiplying and dividing by the unperturbed configurational partition function, Z_0^c,

$$Z_0^c = \int \exp[-\beta V_0(\mathbf{r}^N)] d\mathbf{r}^N \tag{56}$$

The result is[170b]

$$Z_\lambda^c = Z_0^c \langle \exp[-\beta \lambda V_\lambda(\mathbf{r}^N)] \rangle_0 \tag{57}$$

where

$$\langle \exp[-\beta \lambda V_\lambda(\mathbf{r}^N)] \rangle_0 = \frac{\int \exp[-\beta V_0(\mathbf{r}^N)] \exp[-\beta \lambda V_\lambda(\mathbf{r}^N)] d\mathbf{r}^N}{\int \exp[-\beta V_0(\mathbf{r}^N)] d\mathbf{r}^N} \tag{58}$$

Equation 58 can be interpreted as the Boltzmann factor for the perturbation averaged over the unperturbed (reference) system; this is indicated in the expression on the left-hand side of the equation by the angular brackets with the subscript 0. Introducing Eq. 57 into Eq. 53, we obtain

$$\beta \, \Delta A_\lambda = \beta(A_\lambda - A_0) = -\ln \langle \exp[-\beta \lambda V_\lambda(\mathbf{r}^N)] \rangle_0 \tag{59}$$

where A_0 is the unperturbed Helmholtz free energy. Equation 59, which is exact, is the fundamental equation of thermodynamic "perturbation" theory. The change in the Helmholtz free energy as a function of λ is expressed in terms of the exponentially weighted perturbation $\lambda V_\lambda(\mathbf{r}^N)$ averaged over the unperturbed system.

In cases where $V_\lambda(\mathbf{r}^N)$ is small (a perturbation in the true sense), it is possible to expand the exponential involving the perturbation in powers of λ and obtain the simple first-order result,

$$\Delta A_\lambda = \lambda \langle V_\lambda(\mathbf{r}^N) \rangle_0 \tag{60}$$

that is, the free-energy change is given by the perturbation potential averaged over the unperturbed distribution.

A slightly different, though closely related formula can be obtained from the general expression for A_λ (Eq. 53) by differentiating with respect to λ. We find

$$\frac{\partial(\beta A_\lambda)}{\partial \lambda} = \frac{\beta}{Z_\lambda^c} \int \exp[-\beta V(\mathbf{r}^N; \lambda)] \frac{\partial V(\mathbf{r}^N; \lambda)}{\partial \lambda} d\mathbf{r}^N \tag{61}$$

or

$$\frac{\partial (\beta A_\lambda)}{\partial \lambda} = \beta \left\langle \frac{\partial V(\mathbf{r}^N; \lambda)}{\partial \lambda} \right\rangle_\lambda \qquad (62)$$

where the subscript λ means that the derivative of the potential with respect to λ is averaged over the *perturbed* distribution corresponding to the parameter λ. To obtain the free-energy difference between the unperturbed ($\lambda = 0$) and perturbed system (λ), Eq. 62 is integrated with respect to λ to obtain

$$(A_\lambda - A_0) = \int_0^\lambda \left\langle \frac{\partial V(\mathbf{r}^N; \lambda')}{\partial \lambda'} \right\rangle_{\lambda'} d\lambda' \qquad (63)$$

Equation 63 is equivalent to Eq. 59, but it should be noted that in Eqs. 61 through 63 we have not made use of the linearity of λ in the expression for $V(\mathbf{r}^N; \lambda)$. Thus Eq. 63 is valid in the general case of nonlinear perturbative coupling. When the perturbation has the linear form given in Eq. 52, Eq. 63 reduces to

$$\Delta A_\lambda = A_\lambda - A_0 = \int_0^\lambda \langle V_{\lambda'}(\mathbf{r}^N) \rangle_{\lambda'} d\lambda' \qquad (64)$$

Equation 64 shows that ΔA_λ can be calculated by averaging the perturbation potential as a function of λ over the perturbed ensemble. Equation 64 is exact for the linear form of the perturbation, in contrast to Eq. 60 which is a first-order result based on averaging over the unperturbed distribution.

To use either Eq. 59 or Eq. 64 requires ensemble averages of the perturbation potential, $V_\lambda(\mathbf{r}^N)$, that can be evaluated by the dynamical simulation methods described in Chapt. IV. Since only equilibrium properties are being determined, the dynamics of the system obtained from the simulation need not be physically meaningful. For example, it may be convenient when performing structural or thermodynamic calculations by classical mechanical simulations to increase the mass of all the hydrogen atoms by a factor of 10.[170c,d] This improves the efficiency of sampling by shifting the highest-frequency motions into the time range where the bulk of the motions occur. Also, in the absence of such high-frequency motions, a larger time step can be used.

It is possible to eliminate all mass effects and all dynamical information in determining the ensemble averages by the use of Monte Carlo simulation procedures. The direct application of such fully stochastic techniques is not common in the field of macromolecular simulations because the presence of stiff internal degrees of freedom (bond lengths and bond angles) requires that a very small step size be used.[171] Some progress has been made in surmounting this problem for simple flexible molecules (e.g., butane) by introducing Monte Carlo moves that are along normal coordinates.[172] A method that combines some of the best ensemble sampling characteristics for thermodynamic integration of both molecular dynamics and Monte Carlo is the Langevin (or Brownian) simulation method described in Chapt. IV.D.

A problem that arises using computer simulations for evaluating the quantities in Eq. 59, 63 is that if the perturbation is "large" (even the change of a hydrogen atom to a methyl group involved in transforming glycine to alanine is a "large perturbation"), the desired averages converge very slowly. One can perform separate simulations and accumulate differences with repeated application of Eq. 59 or by a quadrature integration of Eq. 63 or 64. Also, as already mentioned, importance or umbrella sampling techniques may be used to obtain the desired ensemble averages.[170a,173] Another possibility is to perform a quasi-continuous λ integration by introducing a small increment in λ at each simulation step. One proposed method uses an increment in λ proportional to the sixth power of the elapsed time with the coefficient adjusted so that λ reaches unity in a prespecified number of simulation steps.[174] With such a continuous sampling algorithm, care is required to make certain that equilibrium averages are obtained for each λ; a useful check is to do the simulation in both directions (i.e., $\lambda = 0$ to $\lambda = 1$ and $\lambda = 1$ to $\lambda = 0$).

For some applications it is convenient to recast the results obtained above in terms of the distribution functions for the system. To keep the notation simple we treat a one-component system consisting of N atoms and assume that it can be described by a potential that depends only on the distance between any pair of the particles. Partitioning the total potential energy as in Eq. 52, we can write

$$V(r; \lambda) = V_0(r) + \lambda V_\lambda(r) \tag{65}$$

with $V_\lambda(r)$ independent of λ. Using the one-dimensional equivalent of Eq. 64 and integrating with respect to λ, we obtain[170]

$$\beta \, \Delta A_\lambda = \beta A_\lambda - \beta A_0 = \tfrac{1}{2}\beta N \int_0^\lambda d\lambda \int \rho g_\lambda(r_{12}) V_\lambda(r_{12}) d\mathbf{r}_1 \, d\mathbf{r}_2 \tag{66}$$

where ρ is the number density of the particles and $g_\lambda(r)$ is the pair distribution function for the perturbed system with perturbation potential $\lambda V_\lambda(r)$. This expression is exact for linear perturbations but requires knowledge of the pair correlation function for each value of the coupling parameter λ. Since this can be difficult to calculate, even with the integral equation approach described below, it is sometimes useful to introduce a first-order approximation to the free energy. Expanding $g_\lambda(r)$ around $\lambda = 0$, we can write

$$g_\lambda(r) = g_0(r) + \lambda[\partial g_\lambda(r)/\partial \lambda]_{\lambda=0} + \cdots \tag{67}$$

where $g_0(r)$ is the distribution function for the unperturbed (reference) system. Combining Eqs. 66 and 67 we obtain the first-order result for $\lambda = 1$,

$$\beta \, \Delta A = \beta A_1 - \beta A_0 = \tfrac{1}{2} \beta \rho \int g_0(r_{12}) V_1(r_{12}) d\mathbf{r}_1 \, d\mathbf{r}_2 \tag{68}$$

Equation 68 is equivalent to Eq. 64 for a one-component system described by a pair potential. Extension to the multicomponent case requires a summation over all pairs of the integrals of the potentials with the unperturbed distribution functions; i.e.,

$$\beta \, \Delta A = \tfrac{1}{2} \beta \sum_{ij} \rho_i \int g_{0,ij}(r_{12}) V_{1,ij}(r_{12}) d\mathbf{r}_1 \, d\mathbf{r}_2$$

Equation 68 is referred to as the high-temperature approximation because the potential appears in the integral as the product βV. Thus, the approximation is valid for small perturbations or high temperatures. Higher-order terms can be included, in principle, to obtain more accuracy. However, the derivatives of the distribution functions with respect to λ involve higher-order distribution functions;[170] e.g., the first-order correction in λ to the distribution function involves three- and four-body distribution functions which are usually difficult to obtain. In some cases, the superposition approximation or other approximate expressions for the higher-order distributions have been introduced.[175] However, the first-order result is the one that has been employed in most applications.[176]

The first-order result is of particular interest because it provides an exact upper bound to the free energy of the perturbed system if the properties of the unperturbed (reference) state are known, as in the case of a hard-sphere fluid. This is obtained from the Gibbs-Bogliubov relationship,[177] which has the form (for a one-component system)

$$\beta A_0/N + \tfrac{1}{2} \beta \rho \int g_\lambda(r_{12}) V_\lambda(r_{12}) dr_{12} \leqslant \beta A/N$$

$$\leqslant \beta A_0/N + \tfrac{1}{2} \beta \rho \int g_0(r_{12}) V_\lambda(r_{12}) dr_{12} \tag{69}$$

Equation 69 may be used to adjust the reference-state parameters so as to minimize the right-hand side of the equation, and obtain the best variational upper bound for A. Since the upper bound is given by the first-order expression, it is straightforward to apply. The lower bound is given by the upper limit of the integral in Eq. 66 and is therefore more difficult to evaluate.

Both the general form of thermodynamic perturbation theory and the reduced expressions in terms of distribution functions are being used to determine the free-energy differences between states of interest. For large and complicated structures, only simulation methods are at present capable of providing the necessary information. In simpler cases, integral equation theories of the liquid state[178] can be used at a great reduction in computational effort. Such theories generally deal directly with the distribution function $g(r)$ and the information necessary for determining the free-energy changes by Eqs. 65 to 68 can be obtained in a relatively straightforward manner.[178,179] The integral equation methods are particularly useful in providing information that is difficult to obtain from simulation studies; an example is given by the study of the effect of different forms for the truncation of the pair interactions at large distances on the thermodynamic properties of the system.[180]

To sketch briefly the integral equation methodology,[178,181] we again focus on a one-component atomic system. The direct correlation function, $c(r)$, can be related to the function $g(r) - 1 = h(r)$ by the Ornstein-Zernike equation[170,178]

$$h(r_{12}) = c(r_{12}) + \rho \int h(r_{13})c(r_{32})dr_3 \tag{70}$$

or, in a simpler notation,

$$h = c + \rho c * h \tag{71}$$

where $*$ denotes a convolution. As this equation contains two unknown functions, $c(r)$ and $h(r)$, a second equation, called a closure relation, is required to solve self-consistently for the correlation functions. A number of different approximate relations between $h(r)$ and $c(r)$ are in current use. One such expression is given by the "hypernetted-chain" equation,[164,170] which has the form

$$c(r) = \exp[-\beta V(r) + h(r) - c(r)] - [h(r) - c(r)] - 1 \tag{72a}$$

By linearizing the exponential with respect to $h(r) - c(r)$, one obtains an alternative closure, called the Percus-Yevick equation,[164,170]

$$c(r) = \exp[-\beta V(r)][1 + h(r) - c(r)] - [h(r) - c(r)] - 1 \tag{72b}$$

Both closures have been employed for determining the distribution functions for liquids;[182] the Percus-Yevick equation tends to yield better results for non-polar systems, while the hypernetted-chain equation (with the appropriate renormalization of long-range interactions)[113,183] is found to be more appropriate for polar and ionic liquids.[184]

In solving the coupled set of integral equations, one may begin with a guess, usually the distribution function for a known system that is similar to the system one wishes to solve. The initial guess is improved by iteration until the resulting distribution function is a self-consistent solution to Eq. 71 and either Eq. 72a or 72b. In most cases the free energy is then estimated by first-order perturbation theory (Eq. 68) or the integration over λ (Eq. 66) is performed numerically. However, under certain conditions the techniques of integral equation theory can be used to perform analytically the λ integral in Eq. 66 by finding an exact differential corresponding to the integrand in Eq. 66 or more generally in Eq. 63. It has been demonstrated that with Eqs. 71 and 72a, but not 72b, the free-energy change in Eq. 66 is given by (for $\lambda = 1$, which is not written explicitly)[114,185]

$$\beta \, \Delta A = \rho \int [\tfrac{1}{2} h(r)^2 - c(r) - \tfrac{1}{2} h(r)c(r)]dr \tag{73}$$

It may be shown that this equation without the h^2 term is equivalent to a Gaussian field theory for the free-energy change.[186] In principle, Eq. 73 could be used with experimental data or with data obtained from computer simulations. However, the simulations are likely to present technical difficulties since the first terms of Eq. 73 are essentially zero-wavevector quantities and, therefore, difficult to obtain with sufficient accuracy[114] by techniques other than integral equation approaches.

Applications of Eq. 73 have been made to small molecules such as butane,[114,186] 1,2-dichloroethane,[114] cyclohexane,[188] to positive and negative ions in aqueous solution[169a,189] and to small peptides[115,115a]. However, the practical robustness and numerical stability of the method still have to be explored before reliable calculations on macromolecules can be undertaken.

CHAPTER VI

ATOM AND SIDECHAIN MOTIONS

The fluctuations of atoms and the motions of sidechains in proteins have been examined experimentally and theoretically. In this section we characterize these two types of motions in terms of their amplitudes, time scales, and other properties by describing a series of theoretical studies related to them. Some results on the functional roles of specific atomic and sidechain motions are also presented. Comparisons with experiment are provided, where available; a more detailed analysis of the experimental measurements is given in Chapt. XI.

A. ATOM MOTIONS

The qualitative features of the atomic fluctuations are illustrated in Figs. 11 and 12. Figure 11 is based on a vacuum molecular dynamics simulation of the bovine pancreatic trypsin inhibitor (BPTI), the first protein to be studied by this method.[15] BPTI is a small protein composed of 58 amino acids and 454 heavy atoms. Figure 11 shows the α-carbons, plus the sulfur atoms involved in the three disulfide bonds. The left-hand drawing represents the X-ray structure and the right-hand drawing is an instantaneous picture of the equilibrated structure after 3 ps.[15] The two structures are very similar, but there are small differences throughout. The largest displacements appear in the C-terminal end, which interacts with a neighboring molecule in the crystal, and in the loop in the lower left, which has rather weak interactions with the rest of the molecule. Corresponding fluctuations relative to the X-ray structure would be observed in "snap shots" taken at any other time during the simulation. Figure 12 shows a sequence of computer drawings of the α-carbons and the heme group of the protein myoglobin based on a 300-ps molecular dynamics simulation.[43,190] The general structure of the protein, including that of the α-helices, is preserved throughout the simulation. However, in each snapshot significant displacements of the main-chain atoms are seen to occur. The heme group, and particularly the proprionic acid groups attached to it, show sizable fluctuations. The nature of the atomic fluctuations in these two examples is characteristic of what is found generally in molecular dynamics simulations of proteins.

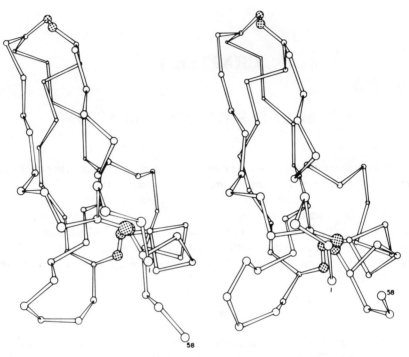

Figure 11. Drawing of α-carbon skeleton plus S—S bonds of PTI; left-hand drawing is the X-ray structure and right-hand drawing is a typical "snapshot" during the simulation. (From Ref. 15.)

1. Amplitudes and Distributions

A number of different proteins, as well as the same proteins with somewhat different potentials and methodologies, have been simulated by molecular dynamics; simulations have been done in vacuum, in solution, and in a crystal environment.[22,42] The results are in general agreement concerning the overall magnitudes and time scales of the atomic fluctuations. Differences that are found are likely to be characteristic of the individual proteins, although some may arise from errors, statistical and otherwise, in the simulations. The root-mean-square fluctuations averaged over all protein heavy atoms (C, N, O, S) are in the range 0.40 to 0.70 Å; backbone atoms tend to have smaller fluctuations (0.30 to 0.60 Å) and sidechain atoms tend to have larger fluctuations (0.50 to 0.90 Å). There is an increase in the magnitude of the fluctuations as one goes from the center of the protein out toward the surface both in vacuum and solution simulations, with significantly larger values for surface residues.

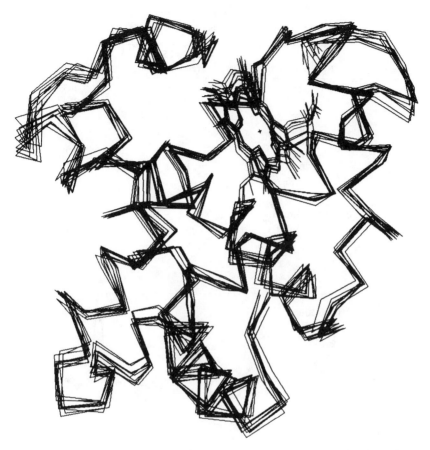

Figure 12. Drawing of myoglobin backbone and the heme group from a series of snapshots at 5-ps intervals in a 50-ps simulation; only the α carbons and the heme group are shown.

Also, sidechain atoms normally increase in their fluctuation amplitudes as one goes away from the mainchain; exceptions can occur, for example, for polar sidechains that are hydrogen bonded, so that the middle portion of the sidechain has larger fluctuations than either end. Table II uses a lysozyme simulation to illustrate some of these results.[191]

Although the average values tend to be similar for different proteins, there is a wide range of variation in the atomic fluctuations within a single protein; i.e., the proteins studied have been found to be inhomogeneous both structurally and dynamically with some regions considerably more flexible than others. Figure 13 shows the variation in the fluctuations of the backbone atoms

TABLE II
Root-Mean-Square Fluctuations of Lysozyme Atoms

Atom Type[a]	RMS Fluctuation	Spherical Shell (radii in Å)	RMS Fluctuation[b]
All	0.76	0–6	0.44
N	0.55	6–9	0.49
C	0.56	9–12	0.53
O	0.75	12–15	0.62
C^α	0.57	15–18	0.73
C^β	0.64	18–21	0.91
γ	0.80	21–14	0.91
δ	0.95		
ϵ	0.95		

[a] Sidechains do not include prolines.
[b] Backbone values only.

Source: Ref. 191

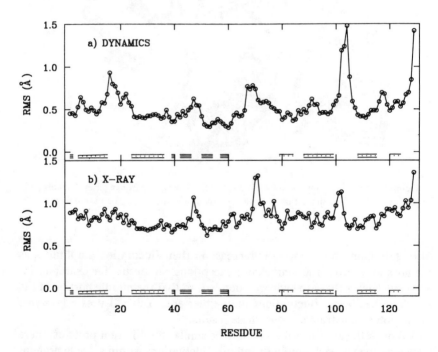

Figure 13. Calculated and experimental rms fluctuations of lysozyme. Backbone averages are shown as a function of residue number, and were obtained (a) from a molecular dynamics simulation and (b) from X-ray temperature factors without correcting for disorder contributions. (From Ref. 192.)

as a function of residue number in a simulation of the protein lysozyme.[192] The hydrogen-bonded secondary structural elements (α-helices and β-sheets) are indicated underneath the figure; for comparison a drawing of the main-chain of lysozyme is shown in Fig. 14. It is evident that the secondary structural elements have smaller fluctuations than the random coil (loop regions) of the protein. When a comparison of different lysozyme crystal structures is made (e.g., two different crystal forms of hen egg white lysozyme or hen egg white and human lysozyme, which are highly homologous),[191] there is a good correlation between regions of the protein that have different conformations in the various structures and those that have large fluctuations in the simulations. Comparison of a simulation of lysozyme with and without an inhibitor bound in the active site showed significant differences in the residue mobili-

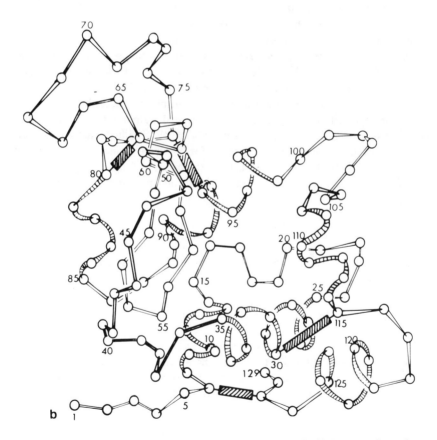

Figure 14. Schematic drawing of lysozyme, based on a sketch by Sir Lawrence Bragg from the original X-ray results. (Courtesy of D. C. Phillips.)

ties. Some of the residues showing differences are in direct contact with the inhibitor, but others are in a region distant from the active site.[192]

Since many studies of proteins have assumed that only rotation about the dihedral angles is significant, the importance of bond-length and bond-angle fluctuations for the atomic motions has been examined. Simulations were performed on BPTI with the bond lengths or both the bond lengths and the bond angles fixed at their average values by use of the SHAKE methodology (see Chapt. IV.G.).[193] Although fixing the bond lengths had no significant effect on the atomic fluctuation amplitude, fixed bond angles (normal fluctuations of $\pm 5°$) reduced the mean amplitude of the atomic motions by a factor of 2. This result demonstrates that in a closely packed system, such as a protein in its native conformation, the excluded volume effects of repulsive van der Waals interactions introduce a strong coupling between the degrees of freedom associated with the dihedral angles and the bond angles.

Of interest also are the results concerning deviations of the atomic fluctuations from simple isotropic and harmonic motion. As discussed in Chapt. XI, most X-ray refinements of proteins assume (out of necessity, because of the limited data set) that the motions are isotropic and harmonic. Simulations have shown that the fluctuations of protein atoms are highly anisotropic and for some atoms, strongly anharmonic. The anisotropy and anharmonicity of the atomic distribution functions in molecular dynamics simulations of proteins have been studied in considerable detail.[193-197] To illustrate these aspects of the motions, we present some results for lysozyme[196] and myoglobin.[197] If U_x, U_y, and U_z are the fluctuations from the mean positions along the principal X, Y, and Z axes for the motion of a given atom and the mean-square fluctuations are

$$\sigma_x^2 = \langle U_x^2 \rangle \qquad \sigma_y^2 = \langle U_y^2 \rangle \qquad \sigma_z^2 = \langle U_z^2 \rangle \qquad (74)$$

with $\sigma_x \geq \sigma_y \geq \sigma_z$, by definition, a measure of the anisotropy is

$$A_1 = \left[\frac{\sigma_x^2}{1/2(\sigma_y^2 + \sigma_z^2)} \right]^{1/2} - 1.0 \qquad (75)$$

The quantity A_1 determines the amount by which the ratio of the fluctuation in the principal X direction to that of the average of the fluctuations in the other two directions (Y and Z) exceeds that of an isotropic distribution, for which A_1 is zero. A second measure of the anisotropy is

$$A_2 = \left[\frac{\sigma_y^2}{1/2(\sigma_y^2 + \sigma_z^2)} \right]^{1/2} - 1.0 \qquad (76)$$

which determines how anisotropic the motion is in the principal Y-Z plane. Both A_1 and A_2 have been calculated for various classes of atoms in lysozyme and myoglobin (see Table III). While the anisotropies corresponding to A_1 or A_2 are slightly lower in myoglobin, the general trends are the same in both molecules. The motions are highly anisotropic in terms of A_1; e.g., in lysozyme, very few atoms (about 1.4%) have A_1 less than 0.02, 61% of the atoms have A_1 greater than 0.5, and 31% have A_1 greater than 0.75. Atoms farther out along the sidechains have higher values of A_1. By contrast, the value of A_2 remains uniformly low for all classes of atoms (at about 0.15). This indicates that the most significant contribution to anisotropy is along the direction of largest motion (the X direction) and that the motion is more isotropic in the Y-Z plane. It is sometimes possible to rationalize this anisotropy in terms of local bonding, e.g., the torsional oscillation of a small group around a single bond.[197] In most cases, however, the directional preferences appear to be determined by larger-scale collective motions involving the atom and its neighbors.[197-199]

The fluctuations of a significant fraction of the protein atoms are found to be anharmonic; i.e., the potentials of mean force for the atomic displacements deviate from the simple parabolic form that would be obtained at sufficiently low temperature. The third and fourth moments of the distribution

TABLE III
Statistics on Anisotropy (A_1) of Atomic Motions[a]

	Myoglobin[b]	Lysozyme[c]
All atoms	0.68 (0.39)	0.85 (0.55)
Backbone	0.57 (0.28)	0.77 (0.50)
Sidechain	0.74 (0.43)	0.93 (0.59)
N	0.55 (0.26)	0.68 (0.30)
C	0.58 (0.28)	0.76 (0.45)
O	0.70 (0.40)	0.93 (0.60)
C^α	0.59 (0.30)	0.73 (0.47)
C^β	0.67 (0.40)	0.74 (0.45)
γ	0.72 (0.46)	0.90 (0.55)
δ	0.76 (0.45)	0.95 (0.56)
ϵ	0.85 (0.47)	1.03 (0.67)
ζ	0.77 (0.42)	1.14 (0.73)

[a] The numbers are averages over all atoms for a particular class, except that proline residues were excluded for the sidechain averages. Numbers in parentheses are standard deviations.
[b] From Ref. 197.
[c] From Ref. 196.

can be used to characterize the anharmonicity.[193,195,196] The skewness, α_{3i}, where i corresponds to the principal axis, X, Y, or Z is defined by

$$\alpha_{3i} = \frac{\langle U_i^3 \rangle}{\langle U_i^2 \rangle^{3/2}} \tag{77}$$

and the coefficient of excess kurtosis, α_{4i}, is given by

$$\alpha_{4i} = \frac{\langle U_i^4 \rangle}{\langle U_i^2 \rangle^2} - 3.0 \tag{78}$$

Both α_3 and α_4 are zero for a Gaussian distribution. Average values of $|\alpha_{3i}|$ and $|\alpha_{4i}|$ for various classes of atoms have been calculated for lysozyme[196] and myoglobin.[197] The results for these two proteins are strikingly similar (Table IV). From an analysis of the atomic distributions, it is apparent that most atoms with large anharmonicity have multiple peaks in their distribution functions, with each peak approximately harmonic. This suggests that the best description of anharmonicity for atoms with large fluctuations cannot be based on a perturbed Gaussian distribution or quasi-harmonic model. Instead, several Gaussians, centered at different positions, could be used to obtain a more accurate description of the distributions.

The effects of anisotropy and anharmonicity in the atomic motions on the refinement of X-ray data for protein crystals are described in Chapt. XI.

TABLE IV
Statistics on Skewness and Kurtosis by Atom Type for Myoglobin and Lysozyme[a]

	U_x	U_y	U_z
A. Skewness $|\alpha_3|$ by Atom Type for Myoglobin[b]			
All atoms	0.38 (0.32)	0.28 (0.25)	0.21 (0.21)
Backbone	0.36 (0.28)	0.26 (0.24)	0.21 (0.17)
Sidechain	0.40 (0.34)	0.29 (0.26)	0.22 (0.24)
N	0.36 (0.27)	0.25 (0.22)	0.20 (0.18)
C	0.37 (0.28)	0.26 (0.25)	0.22 (0.17)
O	0.41 (0.34)	0.26 (0.20)	0.22 (0.17)
C^α	0.35 (0.28)	0.27 (0.26)	0.21 (0.17)
C^β	0.34 (0.30)	0.33 (0.28)	0.20 (0.16)
γ	0.40 (0.37)	0.30 (0.25)	0.23 (0.42)
δ	0.38 (0.34)	0.26 (0.25)	0.21 (0.18)
ϵ	0.40 (0.35)	0.33 (0.31)	0.24 (0.20)
ζ	0.43 (0.34)	0.31 (0.30)	0.22 (0.16)

TABLE IV—*Continued.*

	U_x	U_y	U_z
	B. Skewness $\lvert\alpha_3\rvert$ by Atom Type for Lysozyme[c]		
All	0.38 (0.32)	0.25 (0.23)	0.18 (0.16)
Backbone	0.34 (0.28)	0.22 (0.20)	0.17 (0.14)
Sidechain	0.42 (0.36)	0.28 (0.26)	0.20 (0.18)
N	0.30 (0.25)	0.21 (0.16)	0.16 (0.11)
C	0.33 (0.27)	0.22 (0.21)	0.17 (0.13)
O	0.38 (0.33)	0.27 (0.25)	0.18 (0.15)
C^α	0.33 (0.25)	0.19 (0.17)	0.18 (0.15)
C^β	0.32 (0.24)	0.24 (0.22)	0.17 (0.14)
γ	0.40 (0.36)	0.25 (0.20)	0.18 (0.15)
δ	0.45 (0.38)	0.31 (0.30)	0.21 (0.22)
ϵ	0.53 (0.51)	0.32 (0.27)	0.22 (0.20)
ζ	0.47 (0.36)	0.30 (0.27)	0.21 (0.15)
	C. Kurtosis $\lvert\alpha_4\rvert$ by Atom Type for Myoglobin[b]		
All atoms	0.58 (0.58)	0.45 (0.46)	0.36 (0.67)
Backbone	0.56 (0.46)	0.43 (0.36)	0.36 (0.33)
Sidechain	0.59 (0.64)	0.46 (0.51)	0.37 (0.80)
N	0.53 (0.51)	0.42 (0.33)	0.36 (0.44)
C	0.55 (0.42)	0.45 (0.38)	0.34 (0.25)
O	0.56 (0.67)	0.42 (0.35)	0.31 (0.29)
C^α	0.58 (0.44)	0.44 (0.36)	0.37 (0.27)
C^β	0.49 (0.46)	0.52 (0.61)	0.36 (0.47)
γ	0.60 (0.75)	0.43 (0.44)	0.48 (1.75)
δ	0.63 (0.56)	0.46 (0.48)	0.35 (0.28)
ϵ	0.68 (0.71)	0.52 (0.69)	0.34 (0.27)
ζ	0.72 (0.72)	0.53 (0.53)	0.33 (0.34)
	D. Kurtosis $\lvert\alpha_4\rvert$ by Atom Type for Lysozyme[c]		
All	0.56 (0.52)	0.39 (0.49)	0.31 (0.36)
Backbone	0.50 (0.43)	0.33 (0.37)	0.27 (0.25)
Sidechain	0.61 (0.59)	0.46 (0.58)	0.35 (0.44)
N	0.46 (0.38)	0.30 (0.24)	0.24 (0.19)
C	0.48 (0.38)	0.33 (0.33)	0.26 (0.22)
O	0.57 (0.56)	0.37 (0.56)	0.27 (0.21)
C^α	0.50 (0.38)	0.32 (0.27)	0.31 (0.36)
C^β	0.48 (0.37)	0.36 (0.36)	0.28 (0.20)
γ	0.57 (0.48)	0.42 (0.39)	0.26 (0.25)
δ	0.65 (0.53)	0.54 (0.79)	0.45 (0.72)
ϵ	0.85 (1.07)	0.61 (0.72)	0.46 (0.51)
ζ	0.67 (0.52)	0.40 (0.52)	0.41 (0.24)

[a] Numbers are averages over all the atoms of a particular class, except that proline residues were not included in sidechain averages. Numbers in parentheses are standard deviations.

[b] From Ref. 197.

[c] From Ref. 196.

2. Time Dependence: Local and Collective Effects

The time development of the atomic fluctuations has been examined in BPTI[199] and cytochrome c.[198] To analyze the time dependence for BPTI, a 25-ps trajectory[199] was used and subaveraged root-mean-square (rms) fluctuations were calculated; i.e., the entire trajectory was divided into a series of intervals of given lengths, the mean-square fluctuations relative to the mean positions for each interval were determined and then were averaged for the entire trajectory. Figure 15 shows the results for the C^α atoms. There is a significant contribution to the rms fluctuations from the subpicosecond mo-

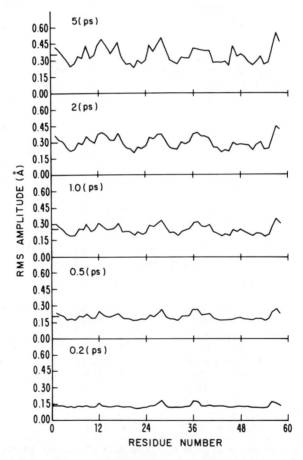

Figure 15. Root-mean-square displacement subaverages as a function of residue number for the C^α atoms ($T_{AV} = 0.2, 0.5, 1, 2,$ and 5 ps).

tions; e.g., at 0.2 ps, the C^α rms value averaged over all residues is already 0.13 Å, about 40% of the result at 5.0 ps. As to the relative values of the C^α fluctuations, the results at 0.2 ps are rather uniform. Since high-frequency oscillations are making the main contribution to this subaverage, it appears that the local effective potential for C^α does not vary significantly throughout the protein. This is in accord with expectations if the dominant factor determining the high-frequency oscillations is the librational potential associated with the torsional motion of the backbone atoms. Even on the 0.2-ps time scale, however, there is some suggestion of inhomogeneity; i.e., in the neighborhood of atoms 12, 28, and 36, as well as of the N- and C-terminal ends, slightly larger fluctuations occur. These regions all have greater-than-average fluctuations when the lower-frequency contributions are included, as can be seen from a comparison of the 0.2- and 5.0-ps values. From the behavior of the time series (Fig. 16), it appears that the 0.5-ps subaverages include all the high-frequency contributions and that the lower-frequency motions are becoming more important. This corresponds to Fig. 15, in that there is a greater variation in the fluctuations along the polypeptide chain at 0.5 ps than at 0.2 ps. As longer time subaverages are examined, the magnitudes of the fluctuations are found to increase in certain regions, in accord with the longer relaxation times and lower-frequency collective character of the motions in-

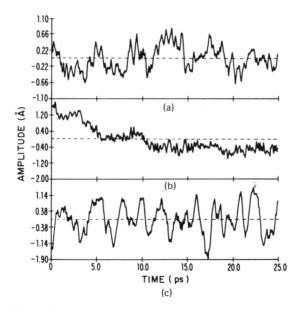

Figure 16. Time series for largest Cartesian component in principal-axis system of thermal ellipsoid: (a) Tyr-21 $C^{\delta 2}$; (b) Asp-50 C^β; (c) Lys-15 C^δ.

volved; examples are the loop region (residues 24 to 28) and the region at the
top of the molecule (around residues 14 and 38) (see Fig. 17). For certain
parts of the protein (e.g., the β-sheet region, 18 to 28), the C^α fluctuations
have already reached their asymptotic values by 2 ps. However, even at 25 ps,
the length of the simulation, asymptotic values for the average rms fluctuations have not been reached for all atoms.

The time series for, and the time development of, the mean-square dis-

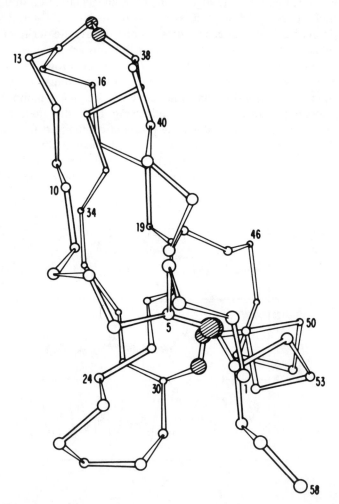

Figure 17. X-ray structure of BPTI with residue numbers; only α carbons and disulfide
bridges are shown.

placements, as well as the correlation functions for the atoms of BPTI,[199] show that the fluctuations generally involve the superposition of two types of motions. One is a high-frequency oscillation of relatively small magnitude, while the other is of considerably lower frequency and larger amplitude. From the characteristics of the individual atomic fluctuations, and from the relationships among the displacements of certain atoms, qualitative conclusions can be drawn concerning the nature of the two types of motional contributions. The high-frequency oscillations are local, in the sense that they correspond to the librational motion of individual atoms in effective potentials. This potential is a summation of dihedral-angle terms in the potential function for the backbone or sidechain of which the atom is a part and of nonbonded interactions with the atoms of the surrounding protein matrix. By contrast, the lower-frequency components have a nonlocal, more collective character, in that they involve the correlated motion of groups of atoms, ranging from a small number of atoms next to each other along the backbone and/or along a single sidechain to a much larger number of atoms (up to 100 or so in some cases) in certain regions of the protein. The frequencies for the collective contributions to the correlation functions vary from 1 to 10 ps^{-1}; in wavenumbers this corresponds to 3 to 30 cm^{-1}. This range is in accord with the lowest normal-mode frequencies obtained for BPTI (see below).

3. Harmonic Dynamics

Early evidence for motion in the interior of proteins or their fragments came from infrared vibrational spectroscopy.[36] It is usually assumed in interpreting such data that a harmonic potential and the resulting normal-mode description of the motions is adequate (see Chapt. IV.F).[200,201] Although it is now known that this approximation is not generally applicable to the atomic motions in proteins (see above), the normal mode description is nevertheless useful for understanding certain aspects of the dynamics. It is most likely to be correct for the mainchain atoms of tightly bonded secondary structural elements, like α-helices and β-sheets.

By using the methods of harmonic analysis, the internal fluctuations of a finite α-helix (hexadecaglycine) were determined from the normal modes of the system.[134] At 300 K, the rms fluctuations of mainchain dihedral angles (ϕ and ψ) about their equilibrium values were equal to about 12° in the middle of the helix and were somewhat larger near the ends. The dihedral angle fluctuations are significantly correlated over two neighboring residues in such a way as to localize the fluctuations.[15,134,200,201] Fluctuations in the distances between adjacent residues (defined as the projection onto the helix axis of the vector connecting the centers of mass of the residues) ranged from about 0.25 Å in the middle of the helix to about 0.5 Å at the ends. The length fluctuations are negatively correlated for residue pairs ($i - 1, i$) and ($i, i + 1$)

so as to preserve the overall length and conformation of the helix; positive correlations are observed for the pairs (4, 5), (8, 9) and (8, 9) (12, 13), suggesting that the motion of residue 8 is coupled to the motions of residues 4 and 12 to retain optimal hydrogen bonding.

Full molecular-dynamics simulations over the temperature range between 5 and 300 K have been performed [137] for a decaglycine helix and the results have been compared with those obtained in the harmonic approximation.[134] For the mean-square positional fluctuations, $\langle \Delta r^2 \rangle$, of the atoms, the harmonic approximation is in good agreement with the molecular dynamics results below 100 K, but there are significant deviations above that temperature; e.g., at 300 K, the average value of $\langle \Delta r^2 \rangle$ obtained for the α-carbons from the full dynamics is more than twice that found in the harmonic model. Quantum effects on the fluctuations are found to be significant only below 50 K. The temperature dependence of the calculated fluctuations is similar to that obtained from X-ray temperature factors for the α-helices in myoglobin between 80 and 300 K.[202]

More recently, harmonic analyses have been extended to proteins. The first calculation was made for BPTI[136] so as to be able to compare the harmonic results with those obtained from molecular dynamics simulations. To avoid approximations, other than those inherent in the harmonic model, the empirical potential function (Eq. 6) employed for dynamics was used to calculate the force constant matrix, and the normal-mode determination was performed in the full conformational space of the molecule; that is, all bond lengths and angles, as well as dihedral angles, were included for the 580-atom system consisting of the heavy atoms and polar hydrogens. Figure 18 shows the normal-mode spectrum of BPTI; Fig. 18a presents all the frequencies up to 2000 cm^{-1} (hydrogen stretching frequencies, which are in the neighborhood of 3000 cm^{-1}, are not shown), Fig. 18b gives the cumulative distribution for the number of modes below a given frequency, and Fig. 18c shows an expanded cumulative distribution for the lowest 300 modes. There is an essentially continuous, although not uniform, distribution of frequencies between 3.1 and 1200 cm^{-1}. In the range 1200 to 1800 cm^{-1}, the frequencies tend to come in groups, many of which are dominated by bond stretching vibrations. There are 20 modes between 3.1 and 13 cm^{-1} and there is a peak in the frequency distribution near 50 cm^{-1}. Other normal-mode calculations have been made for BPTI,[135] as well as for crambin, ribonuclease, and lysozyme,[136a] in a reduced conformational space that included only the dihedral-angle degrees of freedom. Comparison of these results with those from the full calculation shows that, in the former, the frequencies tend to be shifted to higher values and the density of states in the low-frequency region is reduced.

The root-mean-square (rms) atom fluctuations for BPTI were calculated from the normal modes by evaluation of the classical expression given in Eq.

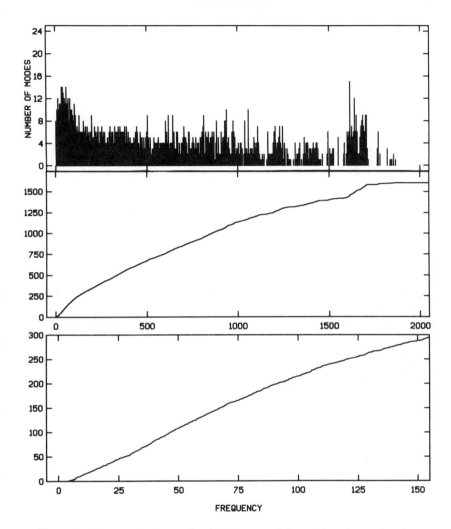

Figure 18. Normal-mode frequencies: (*a*) histogram of the number of normal modes per 5-cm^{-1} interval (hydrogen stretches are not shown); (*b*) number of modes below a given frequency; (*c*) expanded version of (b) in the low-frequency region.

22; as mentioned above, quantum corrections are negligible above 50 K, and, therefore, were not included. Figure 19 shows the normal-mode rms fluctuations calculated at 300 K and compares them with the results of a molecular dynamics simulation of BPTI in a van der Waals solvent.[193] This simulation was used because its average structure is closest to that employed for the normal-mode analysis. The latter was constrained to remain near the X-ray

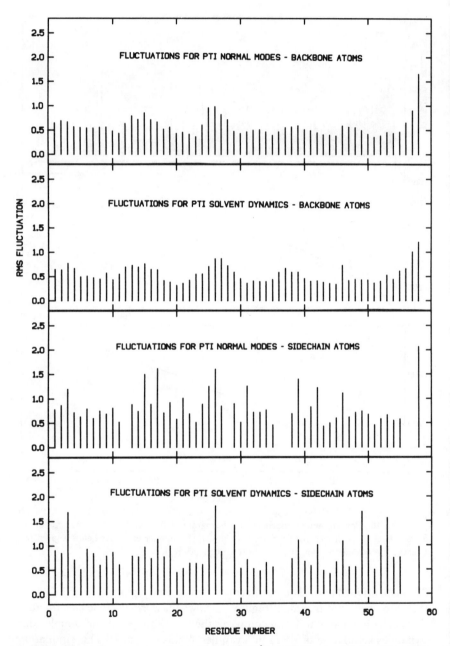

Figure 19. Root-mean-square atomic fluctuations (Å) at 300 K averaged over each residue from normal-mode and molecular dynamics; separate plots are given for mainchain (N, C$^{\alpha}$, C) and sidechain heavy atoms.

structure and so does not correspond to the fully minimized geometry; it is likely that use of a fully minimized structure would result in higher frequencies and smaller fluctuations.

The results for the mainchain and sidechain averages for each residue as a function of residue number are given. For the mainchain fluctuations, the molecular dynamics and normal-mode values are very similar; for the sidechains, there is some correspondence, although the differences are considerably more pronounced. This is in accord with the results on anharmonicity found in the molecular dynamic simulations (see above).

Figure 20 shows the contributions of the different normal modes to the displacements of selected atoms; also shown is the fluctuation of the radius of gyration for the molecule. In most cases, the dominant contributions come from low-frequency modes in the range 3 to 50 cm^{-1}, although nonnegligible

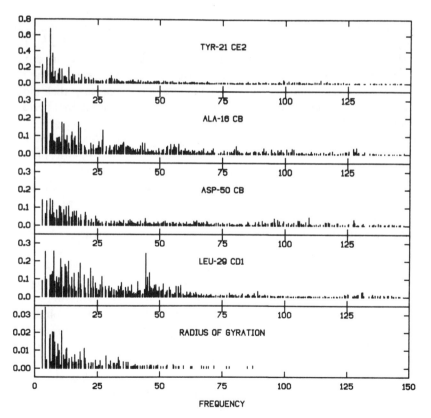

Figure 20. Contribution of normal modes to rms fluctuations (Å) as a function of frequency; selected atoms and the radius of gyration are included.

contributions come from higher frequencies up to 130 cm^{-1}. It is evident that for certain atoms (e.g., Tyr-21 C$^{\epsilon 2}$) only a very small number of modes are important, while for other atoms (e.g., Ala-16 C$^{\beta}$, Asp-50 C$^{\beta}$) a range of frequencies are involved; for Leu-29 C$^{\delta 1}$, a mode at 44.5 cm^{-1} makes a very large contribution. These results for the spectral densities of the rms displacements are in accord with the molecular dynamics analysis; in both calculations the lowest-frequency modes make the largest contributions to the displacements.

The form of the normal modes is of considerable interest. It provides information concerning the correlation between the motions of different groups of atoms. Analysis of the dynamics results (see above) have indicated that the larger-scale motions have a collective character that may involve a few neighboring atoms, a residue, or groups of many atoms in a given region of a protein. In Fig. 21 the distribution of the residue displacements for some of the low-frequency modes are shown; also included is one of the translational modes, which clearly demonstrates the purity of this mode. Most of the 300 lowest modes are delocalized; they are generally distributed over the entire molecule. The lowest mode (at 3.1 cm^{-1}) mirrors the overall mainchain rms fluctuations, as can be seen by comparing this mode in Fig. 21 with Fig. 19. Other modes shown, although they are also delocalized, are distributed differently over the various portions of the molecule. In considering the character of the individual modes, it must be recognized that because of their close spacing, relatively small effects, such as solvent damping or external perturbations (e.g., ligand binding), can lead to significant mode mixing. Thus, rather than individual mode properties, those that involve averages over a range of modes with similar frequencies are likely to be most significant and to be less sensitive to deviations from the simple harmonic model.

From the results of the normal-mode dynamics it is evident that different residues contribute in varying degrees to the different modes of BPTI. This suggests that mutations can affect the internal motions of proteins in specific ways. Thus site-directed mutagenesis may alter not only the structure but also the dynamics of a protein molecule.

Quasi-harmonic simulations of BPTI have also been made.[203,204] One approach[204] made use of a simplified model, in which each residue was represented as a single interaction site,[205,206] and the force constants were estimated from a molecular dynamics simulation.[207] The frequencies obtained were significantly lower than in the harmonic analysis[136] (e.g., the lowest frequency was 0.3 cm^{-1} instead of 3.1 cm^{-1}), presumably because of the approximate model and the neglect of the off-diagonal terms in the force constant matrix. A more realistic quasi-harmonic treatment[204] has also been made with inclusion of all degrees of freedom, except bond stretching, which has been shown not to affect the atomic fluctuations (see above). To calculate the force constant matrix (Chapt. IV.F), a molecular dynamics simulation in

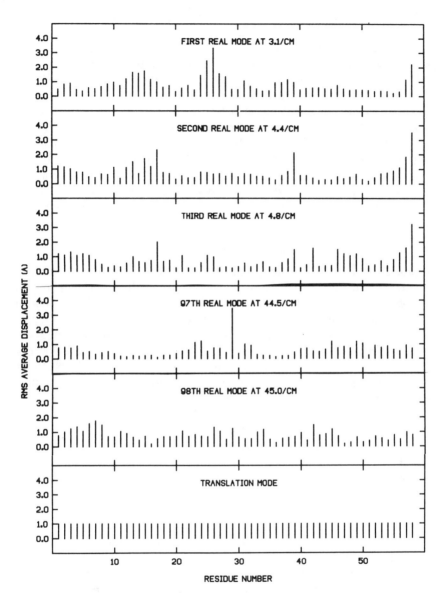

Figure 21. Normal-mode distribution: rms average displacement of atoms within a residue for a 1-Å rms displacement along selected modes.

a van der Waals solvent was used.[193] As expected, the rms atomic fluctuations obtained by averaging over the quasi-harmonic modes are essentially identical to those from the molecular dynamics; if the kinetic energy distribution were fully equilibrated in the latter, identical results would have been obtained. The frequency spectrum found from the quasi-harmonic analysis was very similar to that from the harmonic treatment below 500 cm^{-1}; the lowest quasi-harmonic frequency is equal to 2.7 cm^{-1}, close to the harmonic value of 3.1 cm^{-1}. This correspondence between quasi-harmonic and harmonic results is not surprising in view of the results of the analysis of anharmonicity in proteins from molecular dynamics simulations. In this chapter (Sect. A.1) it was shown that the mainchain fluctuations were rather well approximated by a harmonic model and that where anharmonicity was important, as for certain sidechain atoms, it arose primarily from the presence of multiple conformations.

4.   Biological Role of Atom Fluctuations

Although many of the individual atomic fluctuations observed in the simulations may in themselves not be significant for protein function, they contain information that is of considerable importance. The magnitude of calculated fluctuations demonstrates that the conformational space available to a protein at room temperature includes the range of local structural changes observed on substrate or inhibitor binding for many enzymes. It is possible also that there is a correlated, directional character to the active-site fluctuations in some enzymes that contributes to catalysis. Further, the small-amplitude fluctuations are essential to all other motions in proteins; they serve as the "lubricant" that makes possible larger-scale displacements, such as domain motions (see Table I), on a physiological time scale. In some cases it may be possible to extrapolate from short time fluctuations to larger-scale protein motions.

The larger-scale, collective modes are likely to be of particular significance for biological function. They may be involved, for example, in the displacements of sidechains, loops, or other structural units required for the transition from an inactive to the active configuration of a globular protein. Even if the actual motions have a significant anharmonic character, the normal-mode displacements can serve as a first approximation. Changes in the fluctuations induced by perturbations, such as ligand binding, are also likely to be important. They may result in entropy differences that make a significant contribution to the free energy of binding (see Chapt. X).[136,166a,208]

The extended nature of the collective motions makes them more sensitive to the environment. This is exemplified by the differences in the results between vacuum and solution simulations for BPTI.[193,199] Because such modes involve sizable portions of the protein surface, they could be involved in transmitting external solvent effects to the protein interior.[30] They might also be expected to be quenched at low temperature by freezing of the solvent. Their dominant contribution to the rms atomic fluctuations could explain the transition observed near 200 K in the temperature dependence of these fluctuations in myoglobin[202] and other proteins.

B. SIDECHAIN MOTIONS

The motions of sidechains in proteins play an important role in their dynamics. The time scales involved range from picoseconds for local oscillations in a single potential well to milliseconds or longer for some barrier crossings, such as the 180° rotations (ring "flips") of aromatic sidechains. This range of motions makes it necessary to use a variety of theoretical approaches in the analysis of sidechain dynamics; they include molecular dynamics, activated dynamics, and stochastic dynamics (see Chapt. IV.). There are a number of well-characterized examples where sidechain motions have been shown to play a specific role in protein function.

1. Aromatic Sidechains

The motions of the relatively rigid aromatic sidechains serve as an ideal probe of protein dynamics. As an example, we focus on the tyrosine residues (Fig. 3e) of BPTI. Although their motions are unlikely to have any biological function in this protein, they are of interest for studying the relationship between theory and experiment in a system dominated by short-range nonbonded interactions. Further, historically they represent the first case of a detailed theoretical study of protein motions,[130] which was particularly important in demonstrating, to the considerable surprise of some crystallographers, that 180° rotations of aromatic rings were possible in the interior of a protein.

Before treating the actual dynamics, we consider the nature of the potential experienced by a tyrosine residue in the protein interior. In the native conformation of BPTI, the aromatic ring is surrounded by and has significant nonbonded interactions with atoms of its own backbone and with other residues that are close in space but more distant along the polypeptide chain. Figure 22 shows a potential energy contour map for the sidechain dihedral angles χ^1 and χ^2 of Tyr-21 in a free peptide model (i.e., with the surrounding protein removed), (Fig. 22a) and in the protein BPTI (Fig. 22b). The minimum energy conformations in the two cases are very similar. This appears to be true for most interior residues of proteins; i.e., the observed sidechain conformations are generally close to one of the minima of the isolated peptide.[209] Where the plots differ is that the sidechain is much more rigidly fixed in position by its nonbonded neighbors in the protein than it is by the local interactions in the isolated peptide. Examination of the nonbonded terms shows that

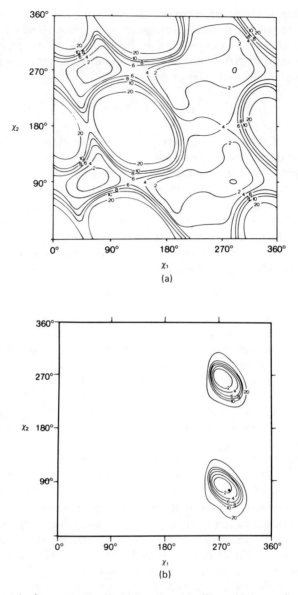

Figure 22. (χ^1, χ^2) maps for Tyr-21: (*a*) free dipeptide; (*b*) peptide in protein; the black dot corresponds to the (χ^1, χ^2) values in the x-ray structure of the protein (ϕ = 253.23°, ψ = 146.77°); energy contours in kcal/mol.

a small number of atoms are involved in constraining the ring in the protein; of particular importance are the backbone N of Ala-48 and $C^{\gamma 2}$ of Thr-32 which are located symmetrically above and below the center of the Tyr-21 ring.

To determine the nature of the fluctuations of aromatic sidechains a room temperature molecular dynamics simulation was analyzed.[153] The results for the relatively buried residue Tyr-21 were compared with those found from a simulation of the dynamics of an isolated tyrosine peptide fragment; for the fragment simulation, the molecular model consisted of BPTI backbone atoms C^{α}_{20} through C^{α}_{22}, together with the Tyr-21 sidechain. The initial coordinates for this fragment were chosen from the BPTI X-ray structure, and the backbone conformation did not vary significantly during the simulation. For both Tyr-21 in BPTI and the Tyr fragment the torsional motions of the ring were examined (i.e., rotations of the ring plane about the $C^{\gamma}_{21} - C^{\zeta}_{21}$ axis), although the axis itself also oscillates.

Figure 23a shows the torsional fluctuations of the Tyr-21 ring observed during the BPTI simulation; the quantity plotted is $\Delta\phi = \phi - \langle\phi\rangle$, where $\langle\phi\rangle$ is the time average of the ring torsional angle. The corresponding torsional fluctuation history for the ring in the tyrosine fragment simulation is shown in Fig. 23b. In BPTI, the root-mean-square fluctuation of the Tyr-21 torsion angle is $12°$, while that for the tyrosine fragment is $15°$. This relatively small difference in amplitudes as compared with that expected from the rigid rotation potential (Fig. 22) makes clear that protein relaxation plays an important role in the ring oscillations. One way of illustrating this is to determine the potential of mean force (see Chapt. IV.D) defined for the angle $\Delta\phi$ by

$$W(\Delta\phi) = -RT \ln P(\Delta\phi) \tag{79}$$

where $P(\Delta\phi)$ is the relative likelihood of a fluctuation $\Delta\phi$ and $P(0)$ is normalized to 1. The potential of mean force obtained from the BPTI simulation is shown in Fig. 24. For comparison we also show the potential $V(\Delta\phi)$ determined by rotating the Tyr-21 ring in the rigid X-ray structure. The potential of mean force, $W(\Delta\phi)$, is significantly softer than the rigid-protein potential, $V(\Delta\phi)$. Correlations between the displacements of the ring and cage atoms in the fluctuating protein thus tend to lower the energy required for a given displacement. The typical cage-atom displacements that contribute to the softening of the potential of mean force are on the order of 0.2 Å, the distance that the δ or ϵ aromatic ring carbons move on a $10°$ torsional rotation.

The comparison in Fig. 23a and b of the dynamics of Tyr-21 in the protein and in an isolated peptide shows that the latter behaves much more like an unhindered oscillator than the former. In the protein the aromatic ring un-

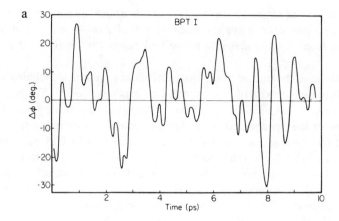

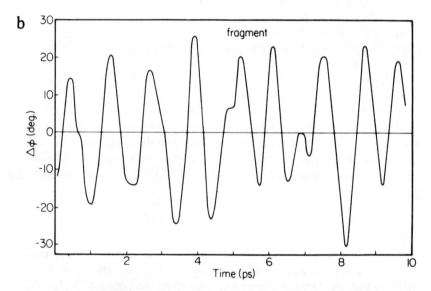

Figure 23. (a) Evolution of the Tyr-21 ring torsional angle during 9.8 ps of dynamical simulation in the protein; (b) evolution of the tyrosine ring torsional angle during 9.8 ps of dynamical simulation of the isolated tyrosine fragment.

dergoes collisions with the surrounding matrix atoms which significantly perturb the motion; some of the interactions are sufficiently strong to reverse the direction of motion, while others produce a smaller change in the angular velocity. As expected, far fewer of these collisional perturbations are evident in the peptide trajectory, for which only interactions with the local backbone are included.

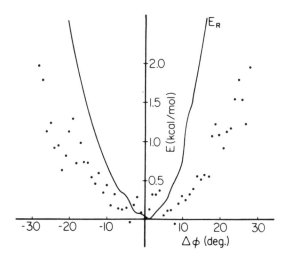

Figure 24. Points represent the potential of mean force for Tyr-21 ring torsional fluctuations, based on statistics from the dynamical simulation. E^R is the total potential energy for Tyr-21 ring torsional displacements in the X-ray structure.

To clarify the dynamic character of the ring fluctuations, it is useful to introduce time correlation functions.[210,211] The time correlation function $C_A(t) = \langle A(s + t)A(s) \rangle$ for a dynamical variable A is obtained by multiplying $A(s)$, the value of A at time s, by $A(s + t)$, the value taken by A after the system has evolved for an additional time, t, and averaging over the initial time s. If the averaging is done over a sufficiently long dynamical simulation of an equilibrated system, $C_A(t)$ will be independent of the initial time, s, used in the calculation; it is then customary to write $C_A(t) = \langle A(t)A(0) \rangle$. If A is the fluctuation of a variable from its mean value, $\langle |A(0)|^2 \rangle$ is the mean-square fluctuation of the variable for an equilibrated system, while the time correlation function, $C_A(t)$, describes the average way in which the fluctuation decays.

The normalized time correlation function for torsional fluctuations of the Tyr-21 ring in BPTI, $C_\phi(t) = \langle \Delta\phi(t) \, \Delta\phi(0) \rangle / \langle |\Delta\phi(0)|^2 \rangle$, is shown in Fig. 25$a$. The torsional oscillations of the ring are seen to be significantly damped, so that the correlation function contrasts sharply with the undamped oscillations expected for an isolated harmonic oscillator; the source of the damping is discussed further below. In Fig. 25b we present for comparison $C_\phi(t)$ calculated from the dynamical simulation of the isolated tyrosine fragment. In this case the tyrosine ring suffers substantially less damping during its torisonal motion; if the integration is continued for longer times, $C_\phi(t)$ exhibits an os-

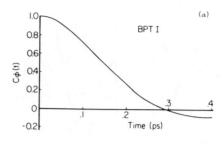

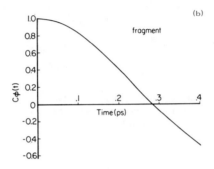

Figure 25. (*a*) Normalized time correlation function for torsional fluctuations of the Tyr-21 ring in the protein. (*b*) Normalized time correlation function for torsional fluctuations of the tyrosine ring in the isolated tyrosine fragment.

cillatory character with only mild damping due to the interactions with the backbone.

The torsional potential of mean force (Fig. 24) and the correlation function for the torsional motions of the Tyr-21 ring in BPTI suggest that the time dependence of $\Delta\phi$ can be described by the Langevin equation for a damped harmonic oscillator (see Chapt. IV.C and D).

$$I\frac{d^2\Delta\phi}{dt^2} + I\beta\frac{d\Delta\phi}{dt} + k\Delta\phi = f(t) \tag{80}$$

Here $I = 7.5 \times 10^{15}$ g-cm^2/mol is the moment of inertia of the ring about the torsional axis, $I\beta$ is the friction constant, k is the harmonic restoring force constant, and $f(t)$ represents the random torques acting on the ring due to fluctuations in its environment. In using the Langevin equation, we implicitly assume that variations in $f(t)$ occur on a much shorter time scale than do variations in $\Delta\phi$; thus $f(t)$ may be regarded as a Gaussian random process and we do not have to specify the details of the mechanism by which the torque fluctuations arise. This time-scale assumption is supported by a collisional model in which it is shown that the mean time between significant changes in $f(t)$ is about 0.07 ps; reference to Fig. 23a shows that on the average, $\Delta\phi$ does not change significantly in this short interval of time. For the restoring force constant, k, it is appropriate to choose a value that includes the effect of cage atom relaxation. A quadratic fit to the potential of mean force $W(\Delta\phi)$, shown in Fig. 24, yields the force constant $k = 5.5 \times 10^{11}$ erg/ (rad^0-mol); the rigid-protein force constant is $k = 2.2 \times 10^{12}$ erg/(rad^2-mol). From Eq. 80 and the definition of relaxation time in the Langevin model, we have $\tau_\phi = I\beta/k$. This time is of the order of 0.2 ps from Fig. 25a, which yields an estimate of the friction constant $I\beta = k\tau_\phi \simeq 0.11$ g-cm²/(s-mol). The ratio $I\beta^2/4k \simeq 0.74$ is less than unity, indicating that the Tyr-21 ring torsional motions are slightly underdamped (i.e., in the absence of random torques, the ring would relax to its equilibrium orientation by damped oscillations).[212]

The friction constant, $I\beta$, may be related to an angular diffusion constant by use of the Einstein formula, $D = k_BT/I\beta$. For the Tyr-21 ring torsional motion in BPTI, one obtains $D = 2.3 \times 10^{11}$ s⁻¹ at 308 K, the temperature of the simulation. This value is somewhat larger than experimental diffusion constants for the corresponding rotational motion of small aromatic molecules in organic solvents (e.g., the value for benzene in isopentane is 8×10^{10} s⁻¹).

The details provided by the molecular dynamics trajectory makes possible a more fundamental analysis of the ring oscillations in the presence of the protein matrix. An approach to the relaxation of fluctuations in gases and liquids, the so-called binary collision model, was pioneered by Enskog[213] and developed by Gordon[214] and Chandler,[215] among others. It pictures the relaxation as occurring as a result of successive binary collisions between the repulsive van der Waals cores of neighboring particles. The successive collisions are assumed to be uncorrelated and instantaneous and to randomize the velocity (or angular velocity) of the struck particles; the particles move freely between collisions. These assumptions are, of course, oversimplifed and modern work on simple liquid dynamics has introduced a number of refinements into the theory.[216] Nevertheless, it appears that Enskog-type models provide a good first approximation. The collisional model used here to describe the Tyr-21 ring torsions takes into account the fact that between collisions, the ring moves in a harmonic manner due to the torsional restoring force. If the collisions are instantaneous and uncorrelated and if each collision randomizes the momentum of the oscillator (in this case the ring angular momentum), the displacement correlation function has the form[153]

$$C_\phi(t) = e^{-\nu t/2}(\cos at + \frac{\nu}{2a} \sin at) \qquad (81)$$

where $a^2 = \omega_0^2 - (\nu^2/4)$ with ω_0 the oscillator frequency and ν the collision frequency, the reciprocal of the mean time between collisions. Equation 81 has exactly the same form as the correlation function obtained from solution of the Langevin equation with the collision frequency ν identified with β. An estimate of the collision frequency from the simulation yields a value of $\nu = 1.4 \times 10^{13}$ s^{-1}, essentially identical with the value of $\beta = 1.47 \times 10^{13}$ s^{-1}. This excellent agreement may be somewhat fortuitous since the basic assumptions of the binary collision model are not satisfied exactly. For example, examination of the effect of the torques on the angular velocity of a tyrosine ring shows that it is not randomized by every collision (Fig. 26).

Activated Dynamics. Although a molecular dynamic simulation provides an excellent approach to the small-angle oscillations of the tyrosine ring, it does not yield information concerning the probability of ring rotations by 180°

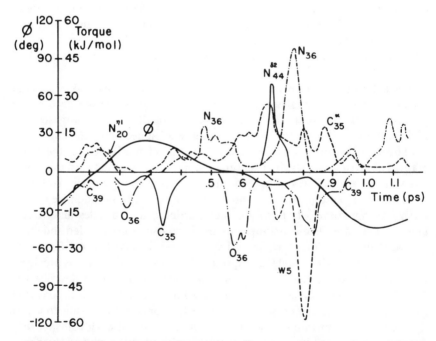

Figure 26. Ring orientation (ϕ) and torques exerted on the ring of Tyr-35 as a function of time as a result of van der Waals interactions with the surrounding protein atoms; the atoms and the associated residues are indicated (contributions that exceeded 0.6 kcal/mol are included).

(ring "flips"). The latter represent a simple example of an activated process in which the rate is limited by an effective energy barrier. Most processes in native proteins that take place on a time scale of nanoseconds or longer involve such an activation step. A standard simulation does not allow one to study activated processes directly because they are, by their nature, rare events; i.e., it is obviously impossible to obtain a statistically valid sample of barrier-crossing trajectories for an activated process with a rate constant of less than 10^{11} s^{-1} in a simulation of length 10 to 100 ps.

In one approach to the 180° rotation problem, which is essentially static in character, empirical energy functions have been used to estimate the activation barriers.[120,130] The method has been applied to BPTI to study the eight aromatic sidechains, four tyrosines, and four phenylalanines. First, the side-chain rotational potentials for the dihedral angle χ^2 were evaluated with the rest of the protein fixed rigidly in position. Since the only atoms displaced are those in the ring, a small portion of the energy function needs to be recalculated for each value of χ^2 and the rotational potential can be determined extremely rapidly. Figure 27 shows the rigid rotation barriers for the ring dihe-

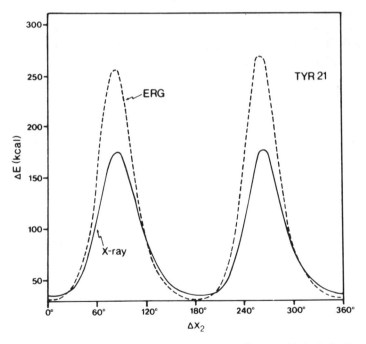

Figure 27. Rigid rotation barriers for the aromatic ring (χ^2) of Tyr-21; both the X-ray and energy-refined geometry (ERG) results are shown.

dral angle χ^2 of Tyr-21. The barriers obtained using the X-ray and an energy-refined geometry are very high. The location of the maximum of the barrier is near 90° and it has an essentially symmetric form.

The rotation barriers found in this way are so large (see Table V) that the protein is not rigid enough to maintain them. Instead, "relaxation" of the protein occurs during the rotation of an aromatic ring and the effective barriers are much reduced. This is due to the fact that the dominant contributions to the barriers come from a small number of repulsive nonbonded contacts. Since such interactions are short range (an r^{-12} distance dependence; see Chapt. III), a small displacement that does not significantly disturb the rest of the protein can lead to a large reduction in the effective barrier. To evaluate the importance of protein relaxation, adiabatic barriers[217,218] were determined. This was done by rotating the dihedral angle χ^2 to a given value and constraining it to that value by introducing a large quadratic term into the potential function. With this constraining potential, all other degrees of freedom in the full conformational space of the protein were allowed to relax in accord with the steepest-descent minimization procedure (Chapt. IV.H). At the minimum ($\Delta\chi^2 = 0°$) and maximum energy orientation of the aromatic ring, 150 steepest-descent cycles of energy minimization were performed starting with the coordinates corresponding to the energy-refined geometry. This number of cycles is sufficient for approximate convergence of the energy difference between the two orientations (i.e., for estimating the barrier height), although both structures are still slowly decreasing in energy. Table V lists the resulting barriers for the four tyrosines of BPTI[130]; Tyr-35 stands out as having the highest barrier.

The Tyr-21 barrier provides a good example of the mechanism by which steric repulsions are lessened as a result of shifts in atom positions. For a 50-cycle minimization, the barrier is abut 15 kcal/mol, of which 11 kcal/mol is

TABLE V
Barriers and Rates for Ring Flips in BPTI

Residue	Rigid Protein Barrier (kcal/mol)	Adiabatic Protein Barrier (kcal/mol)	Theory Rate[a] (s^{-1})
Tyr-10	43	~0	1.89×10^{12}
Tyr-21	230	12	3.4×10^3
Tyr-23	75	7	3.3×10^7
Tyr-35	200	23	3.8×10^{-5}

[a] Rates calculated at 300 K using Eq. (82) with adiabatic barrier, $\nu_T = 8.6 \times 10^{12}$ and $\kappa = 0.22$ (see text).

Source: Ref. 218 and as discussed in the text.

due to nonbonded repulsions; the bond-angle strain is about 5 kcal/mol and all other contributions amount to -1 kcal/mol. Thus the nonbonded part of the barrier has been reduced by about 214 kcal/mol at a cost of 5 kcal/mol in bond-angle strain. The additional 100 cycles reduce the barrier by 3 kcal/mol.

With the barrier results given in Table V it is possible to estimate the rate constant for ring rotation by using analytic models for the dynamics involved. The simplest assumption is to regard the rotation as a unimolecular process that can be treated by transition-state theory with a classical rate constant k equal to

$$k = \kappa \nu_T e^{-E_A/RT} \tag{82}$$

where ν_T is the vibrational frequency at the bottom of the well and κ is the transmission coefficient. The frequency ν_T is estimated to be 8.6×10^{12} s^{-1} from molecular dynamics results.[153] If the transition-state value of κ equal to 1 is used, an upper limit for the reaction rate is obtained.[127] An improved method is to use the formulation of Kramers for a high barrier that takes account of collisional damping by the environment.[219] In this formulation Eq. 82 is modified by introducing an expression for the transmission coefficient κ equal to

$$\kappa = 2\pi \nu_b/\beta \tag{83}$$

where ν_b is the frequency at the top of the inverted barrier, and the value of the friction β can be determined from the results given earlier in this section. As shown below, detailed activated dynamics calculations suggest that κ is equal to 0.22. With this value of κ and E_A set equal to the adiabatic values for the barriers, Eq. 82 gives the rate constant listed in the last column in Table V. It can be seen that all aromatic residues except Tyr-35 are expected to appear to be freely rotating on the time scale of 10^3 s^{-1}, corresponding to the nuclear magnetic resonance measurements, in agreement with the analysis of Snyder et al.[118] and Wagner et al.[119] For Tyr-35, which is in the range for quantitative NMR rate measurements, the calculated rate constant at 300 K is 3.3×10^{-5} s^{-1} to be compared with the experimental value of 0.6 s^{-1}.[119] The fact that the calculated rate constant is smaller than the experimental value is due, in part, to the fact that the adiabatic estimate of the barrier is too high. A value for E_A equal to 17 kcal/mol gives a rate constant in agreement with experiment; an adiabatic barrier of 16 kcal/mol was estimated with the reaction coordinate described below, instead of the rotation about χ^2.[220] A more detailed discussion of the experiments is given in Chapt. XI.B.

Although the static reaction-path studies just described provide an approximate value for the energy barrier, they cannot give information concerning the dynamics of the activated process or of the entropic contribution to the rate, except as estimated from rate-theory models, as in Table V. To overcome the limitations of standard reaction path and molecular dynamics calculations, a synthesis of these techniques with the widely used concepts of transition-state theory can be employed. This is the activated dynamics method described in Chapt. IV.E. In what follows we illustrate the activated dynamics method by applying it to the problem of aromatic ring flips in BPTI.[122-124] We describe the determination of the transition-state region and the reaction coordinate, the evaluation of the probability of being in the transition-state region, the generation of transition-state configurations, and finally the evaluation of the transmission coefficient. For specific study Tyr-35 was chosen; like Tyr-21 it is in the interior of the protein. Because the rotational barrier is dominated by nonbonded interactions between ring atoms and those of the surrounding protein, this case is an excellent example for analysis of the dynamics of a process in which the effects of protein relaxation and frictional damping are expected to be important.

Although the tyrosine ring dihedral angle χ^2 is the obvious reaction coordinate for the rotation, a series of trajectories[123] calculated for different initial configurations of the protein demonstrated that the value of χ^2 at the barrier maximum varied significantly. This means that there are additional protein atoms whose nonbonded interactions with the ring systematically contribute to the barrier. The positions of such atoms must be included in the choice of a suitable reaction coordinate. The trajectory results showed that one of the ring carbon atoms ($C_{35}^{\delta 2}$) and the mainchain nitrogen of the subsequent amino acid (N_{36}) consistently have a large repulsive interaction in the transition-state configurations, whereas the other atoms involved varied with the trajectory. This repulsion arises when one edge of the ring squeezes by the local backbone during rotations in which χ^2 changes from approximately 60 to 240°, the endpoints being the two equivalent minimum-energy regions with the standard convention that the dihedral angle $\chi^2 = 0$ corresponds to a cis configuration. Thus, the coordinate χ^2 was replaced by the reaction coordinate $\xi = \chi^2 - \chi_\nu$, where χ_ν is the virtual dihedral angle $C_{35}^{\beta} - C_{36}^{\gamma} - C_{35}^{\delta 2} - N_{36}$ that characterizes the orientation of the ring plane relative to the local backbone. A more general reaction coordinate could be determined by maximizing the transmission coefficient for motion through the transition state in both directions.[127] However, the results obtained with ξ (see below) indicate that it is a satisfactory approximation to the optimum reaction coordinate.

To evaluate the probability of finding the system in the transition region and from that the free energy or potential of mean force $W(\xi)$ (Chapt. IV.D and E), a series of overlapping umbrella sampling trajectories was performed with harmonic potentials that shifted the equilibrium position of the tyrosine ring along the reaction coordinate from the initial well to the top of the barrier. To simplify the calculation and eliminate large-scale deformation of the protein, only 94 atoms in a restricted region within 7.7 Å of the ring centroid were allowed to move in this primitive version of a boundary simulation. The resulting potential of mean force, $W(\xi)$, evaluated from Eq. 79, and the average potential energy, $\langle V(\xi)\rangle$, as a function of the reaction coordinate ξ for the ring rotation are presented in Fig. 28. Given the model, the statistical errors in the values of $W(\xi)$ and $\langle V(\xi)\rangle$ are estimated to be ± 3 kcal/mol, based on the differences in calculated values of $W(\xi)$ and $\langle V(\xi)\rangle$ for a series of independent sample runs.

The potential of mean force in the transition state, $W(\xi\dagger)$, is the difference in Helmholtz free energy between the top of the barrier and the minimum in

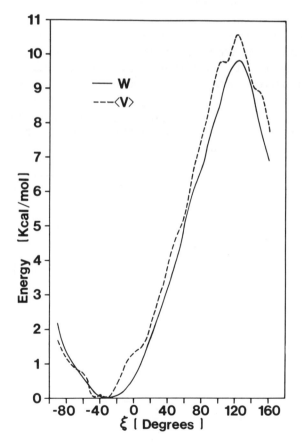

Figure 28. Potential of mean force $W(\xi)$ (solid curve) and the average potential energy $\langle V(\xi)\rangle$ (dashed curve) as functions of the Tyr-35 ring rotation reaction coordinate ξ.

the reactant potential well. We write $W(\xi\dagger) = \Delta E(\xi\dagger) - T\,\Delta S(\xi\dagger)$, where $\Delta E(\xi\dagger)$ and $\Delta S(\xi\dagger)$ are the differences in internal energy and entropy, respectively, between the top of the barrier and the reactant minimum. Because the kinetic energy of the system remains constant at constant temperature, as the value of the reaction coordinate is varied from the initial to the transition state, we have $\Delta E(\xi\dagger) \simeq \langle V(\xi\dagger)\rangle - \langle V(\xi_i)\rangle \equiv \Delta\langle V(\xi\dagger)\rangle$; i.e., the mean potential energy difference, $\Delta\langle V(\xi\dagger)\rangle$, is the effective transition-state energy barrier. By contrast, the quantity, $W(\xi\dagger)$, is the free-energy barrier, which includes the entropy effects arising from relaxation of the surrounding protein matrix. That $W(\xi\dagger)$ and $\Delta E(\xi\dagger)$, are identical to within ± 2 kcal implies that the entropic contribution to the ring-isomerization reaction due to the protein is negligible. This is an important and perhaps somewhat surprising result, particularly in view of the interpretation of NMR measurements for the ring rotation reaction as a function of temperature (see Chapt. XI.B).

The transmission coefficient κ was calculated from a series of transition-state trajectories by monitoring the recrossings ($\xi = \xi\dagger$) that occur as a function of time until each trajectory is finally trapped in the product or reactant well. The normalized reactive flux-correlation function $\kappa(t)$ defined in Eq. 19 was constructed from this set of trajectories;[131] the result is shown in Fig. 29. From its initial value, equal to the transition-state result, $\kappa(t)$ decreases rapidly until it becomes approximately constant for an extended period. The ra-

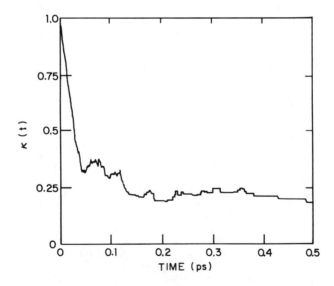

Figure 29. Value of the reactive flux-correlation function $\kappa(t)$ versus time; the function is normalized to the transition-state-theory value (see the text).

tio of this "plateau value" to the initial value of the correlation function is equal to the transmission coefficient κ;[131,131a,221] the transmission coefficient $\kappa \simeq 0.22$. This is the value that was used in calculating the rates of ring flipping shown in Table V.

The fact that the transmission coefficient, κ, is somewhat less than unity can be due to two sources. First, the collisions between the ring and surrounding matrix may be sufficiently rare that recrossings of the transition-state ridge occur before the rotational motion has been quenched in the product valley (corresponding to the low-pressure "falloff" for gas-phase unimolecular reactions).[222] Second, the collisions may be so frequent that many of them occur as the system is crossing the barrier, and the motion across the barrier becomes diffusive; this corresponds to the Kramers limit.[219,223] The calculated trajectories show that the system is not in the high-collision-frequency regime, but in an intermediate region, where κ differs by less than one order of magnitude from the transition-state value of unity. From the trajectories it is clear that damping or frictional effects are present but do not dominate inertial effects in the barrier-crossing dynamics. This qualitative result is in accord with the analysis of the equilibrium torsional fluctuations of tyrosine rings in BPTI (see above). It indicates that transition-state theory, with the appropriate choice of reaction coordinate, is approximately valid for the rotational isomerization of Tyr-35 in the protein interior. This contrasts with stochastic dynamics models of butane isomerization in aqueous solution, for example, where the reaction is close to the fully damped (Kramers) regime.[224]

To obtain a more detailed understanding of the collisional origin of κ, it is useful to look at the torques experienced by the ring during a 180° rotation. Figure 30a shows the time variation of the ring torsional angle and torsional angular velocity for one such case. The torsional motion of the ring was nearly stopped one or more times during the barrier crossing. A detailed analysis shows that the torsional motion of the ring can be accounted for in terms of nonbonded repulsions between ring and protein matrix atoms. More specifically, the total impulse due to the resulting torques that exceed 2 kcal/mol in magnitude at any instant is nearly equal to the observed angular momentum change of the ring during the interval over which these torques act. The time variations of all such torques during the trajectory is shown in Fig. 30b. Examination of this figure and corresponding ones for other trajectories shows that most of these torques have substantial magnitudes only for rather short intervals (≤ 0.1 ps). This suggests that a collisional description is appropriate in first-order analytical models for the barrier-crossing dynamics. The torque impulses are similar to those that occur when the ring oscillates about its equilibrium orientation. The ring is driven over its rotational barrier not as the result of an unusually strong collision, but as the result of a decrease in the frequency and intensity of collisions that would tend to drive the ring away

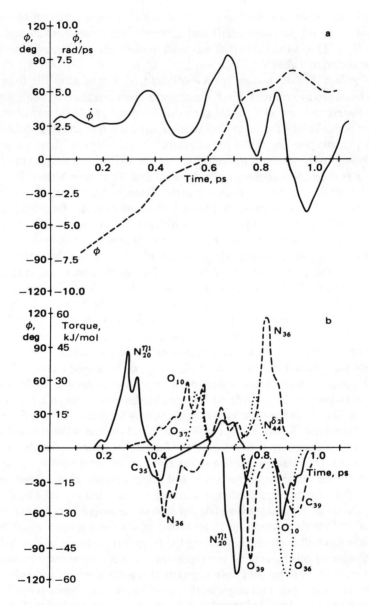

Figure 30. Barrier-crossing trajectory of Tyr-35: (a) ring torsion angle (ϕ) and torsional angular velocity ($\dot{\phi}$) as a function of time; (b) torques exerted on the ring by particular matrix atoms due to nonbonded interactions (contributions from all atoms for which the repulsion exceeded 0.6 kcal/mol are included).

from the barrier. This observation suggests that transient packing defects play a role in initiating the ring rotation. In an analysis of the structural fluctuations accompanying the ring rotation, it has been shown that a portion of the protein backbone above one face of the ring moves away from the ring prior to a successful rotation that leads to the isomerization.[220]

The value of κ, obtained from the reactive flux correlation function can be combined, as in Eq. 18, with the relative probability of being in the transition-state region $[\rho(\xi\dagger)/\int_i \rho(\xi)d\xi = 8.1 \times 10^{-8}\,\text{rad}^{-1}]$ and the mean absolute value of the crossing velocity $(<|\dot{\xi}\dagger|> = 1 \times 10^{13}\,\text{rad/s})$ to obtain a rate constant for the ring flip; the resulting value is $k \simeq 8.9 \times 10^4\,\text{s}^{-1}$ at 295 K. This value is much larger than the simple estimate in Table V due to the smaller activation barrier (9.8 kcal/mol) obtained here, in comparison with the adiabatic value (23 kcal/mol). The resulting rate is also considerably larger than the experimental value; a discussion of this difference is given in Chapt. XI.B.

2. Ligand-Protein Interaction in Myoglobin and Hemoglobin

A biologically significant process in which sidechain motions play an important role is the migration of ligands like carbon monoxide and oxygen from the solution through the protein matrix to the heme group in myoglobin and hemoglobin and then out again. Examination of the high-resolution X-ray structure of myoglobin[225] does not reveal any path by which such ligands can move between the heme binding site and the outside of the protein. Since this holds true both for the unliganded and liganded protein (i.e., myoglobin and oxymyoglobin),[225,226] structural fluctuations must be involved in the entrance and exit of the ligands. Empirical energy function calculations[117] have demonstrated that the rigid protein would have barriers on the order of 100 kcal/mol; such high barriers would make the transitions infinitely long on a biological time scale. Figure 31 shows a potential energy map calculated from the X-ray structure of myoglobin for the shortest "path" from the heme pocket to the exterior of the protein. The figure gives the nonbonded potential energy contour lines for a test particle representing an O_2 molecule interacting with the protein atoms in a plane (xy) parallel to the heme and displaced 3.2 Å from it along the z axis in the direction of the distal histidine; the coordinate system in this and related figures has the iron at the origin and the z axis normal to the heme plane. The low potential energy region in the center is the so-called "heme pocket," with the energy minimum corresponding to the observed position of the distal O atom of an O_2 molecule forming a bent Fe—O—O bond.[225] The shortest path for a ligand from the heme pocket to the exterior (the low-energy region in the upper left of the figure) is between His-E7 and Val-E11. However, this path is not open in the X-ray geometry because the energy barriers due to the surrounding residues indicated in the

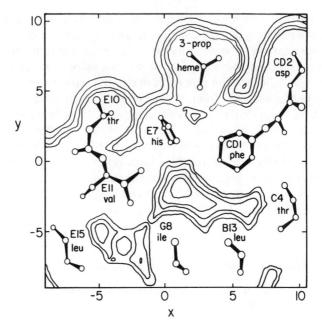

Figure 31. Myoglobin-ligand interaction contour map in the xy plane at $z = 3.2$ Å (see the text). Distances are in Å and contours in kcal; the values shown correspond to 90, 45, 10, 0 and −3 kcal/mol relative to the ligand at infinity. The highest contours are closest to the atoms whose projections onto the plane of the figure are denoted by circles.

figure are greater than 90 kcal/mol. Figure 32 shows a possible path to the exterior in a plane (xy plane) perpendicular to that of Fig. 31; again the barriers in the X-ray structure are very large.

To determine pathways available in the thermally fluctuating protein, ligand trajectories were calculated using the static myoglobin X-ray structure together with a test molecule of reduced effective diameter to compensate in an approximate manner for the absence of protein motions.[117] The trajectory was determined by releasing the "photodissociated" test molecule with substantial kinetic energy (15 kcal/mol) in the heme pocket and following its classical motion for a suitable length of time. A total of 80 such trajectories were computed; a given trajectory was terminated after 3.75 ps if the test molecule had not escaped from the protein. Slightly more than half the test molecules failed to escape from the protein in the allowed time; 15 molecules remained trapped near the heme binding site, while another 21 were trapped in two cavities accessible from the heme pocket. Most of the molecules which escaped did so between the distal histidine (E7) and the sidechains of Thr-E10

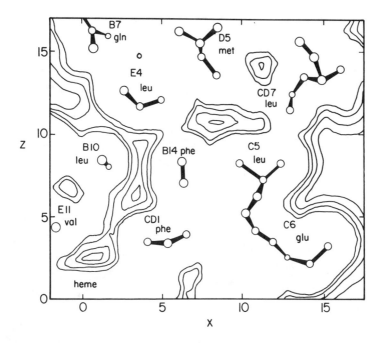

Figure 32. Myoglobin-ligand interaction contour map in the xy plane at $y = 0.5$ Å (see Fig. 31 legend).

and Val-E11 (see Fig. 31). A secondary pathway was also found (see Fig. 32); this involves a more complicated motion along an extension of the heme pocket into a space between Leu-B10, Leu-E4, and Phe-B14, followed by an escape between Leu-E4 and Phe-B14. A typical model trajectory along this path is shown in Fig. 33. More complex pathways also exist, as indicated by the range of motions observed in the trajectories. This has been confirmed recently by simulations in which the protein, as well as the ligand, were allowed to move.[226a]

In the rigid X-ray structure, the two major pathways described above have very high barriers for a ligand of normal size. Thus, it was necessary to study the energetics of barrier relaxation to determine whether either of the pathways had acceptable activation enthalpies. Local dihedral angle rotations of key sidechains, analogous to the tyrosine sidechain motions described above (this chapter, Sect. B.1), were investigated and it was found that the bottleneck on the primary pathway could be relieved at the expense of modest strain in the protein by rigid rotations of the sidechains of His-E7, Val-E11, and Thr-E10. The reorientation of these three sidechains and the resultant open-

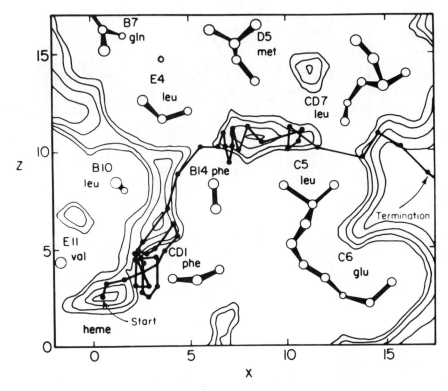

Figure 33. Diabatic ligand trajectory following the secondary pathway (see text); a projection on the plane of Fig. 32 is shown with the dots at 0.15-ps intervals. The start of the trajectory at the heme iron and the termination point exterior to the protein are indicated by arrows.

ing of the pathway to the exterior is illustrated schematically in Fig. 34; panel I shows the X-ray structure (same as Fig. 31); in panel II the distal histidine (E7) has been rotated to $\chi^1 = 220°$ at an energy cost of 3 kcal/mol; in panel III, Val-E11 has also been rotated to $\chi^1 = 60°$ (5 kcal/mol); and panel IV has the additional rotation of Thr-E10 to $\chi^1 = 305°$ (1 kcal/mol). In this manner a direct path to the exterior has been created with a barrier of about 5 kcal/mol at an energy cost to the protein of approximately 8.5 kcal/mol, as compared with the value of nearly 100 kcal/mol calculated for the X-ray structure. Along the secondary path, no simple torsional motions reduced the barrier due to Leu-E4 and Phe-B14, since the necessary rotations led to larger strain energies.

    To determine more directly the energetics and types of motions involved in both the primary and secondary pathways, an adiabatic calculation was per-

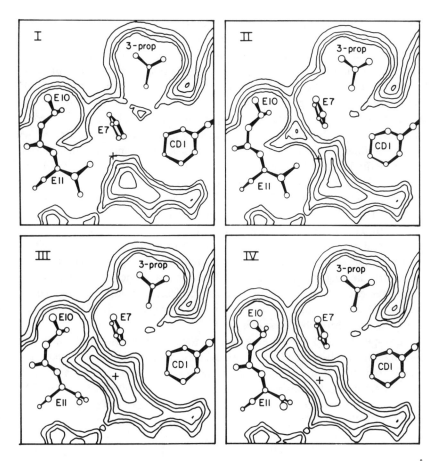

Figure 34. Myoglobin-ligand interaction contour maps in the heme xy plane at $z = 3.2$ Å (the iron is at the origin) showing protein relaxation; a cross marks the iron atom projection onto the plane. Distances are in Å and contours in kcal; the values shown correspond to 90, 45, 10, 0, and -3 kcal/mol relative to the ligand at infinity. The highest contours are closest to the atoms whose projections onto the plane of the figure are denoted by circles. Panel I: X-ray structure. Panels II–IV: sidechain rotations discussed in text.

formed starting with the energy-refined structure of myoglobin. A test sphere with a van der Waals radius of 3.2 Å was fixed at one of the bottlenecks on the primary path (between His-E7 and Val-E11, or between His-E7 and Thr-E10) or in the bottleneck on the secondary path (between Leu-E4 and Phe-B14). The protein was allowed to relax by energy minimization (adiabatic limit) in the presence of the ligand and the resulting displacements in the polypeptide chains were monitored. There were local alterations in sidechain dihedral angles and bond angles. In addition, neighboring sidechains and the backbones of helices D and E participated in the globin response, mostly by small dihedral-angle changes. Approximate values for the relaxed barrier heights were found to be 13 and 6 kcal/mol for the two primary path positions and 18 kcal/mol for the secondary path position. The various barriers calculated by energy minimization in the adiabatic limit are on the order of those estimated by flash photolysis, rebinding studies for carbon monoxide in myoglobin.[227]

A preliminary activated trajectory analysis of the ligand motion,[228] based on the methodology developed for the tyrosine ring rotation analysis (Chapt. IV.E and V.B.1), suggests that one of the barriers along the primary path (namely, that between His-E7 and Thr-E10) is dominated by entropic factors, i.e., that the enthalpic component of the potential of mean force barrier is near zero. This is in disagreement with the adiabatic calculations just described and photolysis experiments.[227] Since protein relaxation is likely to yield enthalpic barriers that are too low in activated dynamics calculations (as in the tyrosine ring rotation study), the temperature dependence of the theoretical free-energy barrier should be obtained from simulations to determine the relative contributions of the enthalpy and entropy of activation to the motion of ligands through the protein matrix.

The type of ligand motion expected for the several-barrier problem found in myoglobin has been evaluated from the trajectory studies.[117,226a] What happens is that the ligand spends a long time in a given well, moving around in and undergoing collisions with the protein walls of the well (see Fig. 33). When there occurs a protein fluctuation sufficient to lower the barrier significantly or the ligand gains sufficient excess energy from collisions with the protein, or most likely both at the same time, the ligand moves rapidly over the barrier and into the next well, where the process is repeated. That the ligand spends most of the time in the low-energy wells is evident from Fig. 33. However, it should be noted that in a realistic trajectory involving a fluctuating protein and ligand-protein energy exchange, the time spent in the wells would be much longer than that found in the diabatic model calculations described here; this is due to the fact that the latter do not give realistic values for the time required to cross high energy barriers. Further, from the complexity and the range of pathways in the protein interior, it appears likely that the motion of the ligand will have a diffusive character.

The analysis of myoglobin and more general considerations of the atomic packing in native proteins suggest that, in many cases, small molecules cannot enter or leave the binding site if the protein atoms are constrained to their average positions. Consequently, sidechain and other fluctuations must be required for ligand binding by many proteins and for the entrance of substrates and exit of products from certain enzymes.

CHAPTER VII

RIGID-BODY MOTIONS

In this chapter we focus on the internal motions of proteins that can be described approximately as displacements of groups of atoms whose relative positions are kept fixed. These include structural changes and fluctuations involving helices, domains, and subunits (see Table I). We present examples for which dynamical as well as structural studies are available. It is clear from the results of such studies that the atomic and sidechain fluctuations (see Chapt. VI) that accompany the rigid-body motions play an important role in reducing the energies involved, and in cases where activation barriers are present, in obtaining rates on a time scale that permits them to be of functional significance.

A. HELIX MOTIONS

For proteins with significant helical content it is possible to show that larger-scale displacements involve rigid-body motions of the helices. Two such proteins for which molecular dynamics simulations have been analyzed in terms of helix motions are the C-terminal fragment of the L7/L12 ribosomal protein of *E. coli*[229] and myoglobin.[230]

The C-terminal fragment of L7/L12 consists of 68 amino acids and is composed of a layer containing three antiparallel α-helices over a twisted antiparallel β-sheet with a β-α-α-β-α-β connectivity. A 150-ps vacuum simulation[229] was performed at 277 K with a polar hydrogen model (i.e., only polar hydrogens were explicitly included in the simulations; see Chapt. III); the first 20 ps was used as the equilibration period and the remainder of the simulation served for analysis. A feature of the simulation was that one of the α-helices, helix B, underwent a librational motion with respect to the rest of the molecule. The interhelical angle between helices B and C fluctuated with a maximum amplitude of approximately 15° and a period of 6.7 ps. Helix C showed considerably less motion and it was overdamped in character. It appears that the connection of helix C at both ends to a strand of relatively rigid β-sheet holds it in place, in contrast to helix B, which is connected by a highly flexible turn to helix A. These results and those from other simulations suggest that varying degrees of rigid-body motions can occur for helices in proteins and

117

that the specific character of the motion is determined by the structure and connectivity.

Additional information concerning helix motions and the more general problem of multiple minima in proteins has been obtained from a 300-ps simulation of myoglobin[190] performed at 300 K with an extended atom model (see Chapt. III). To determine the topography of the potential surface underlying the dynamics, coordinate sets at 10-ps intervals were chosen and subjected to energy minimization.[230] Since all of these structures correspond to separate minima, coordinate sets with smaller time intervals between them were examined; if two coordinate sets converged on minimization in terms of the root mean square (rms) difference between the structures, they were assumed to correspond to the same minimum, while if they diverged they were assumed to correspond to different minima. It was found that the molecule remained in the same minimum for 0.15 ± 0.05 ps. Thus the 300-ps simulation sampled on the order of 2000 minima, a number that is likely to be only a small fraction of the total number of minima accessible in the neighborhood of the native structure to a protein such as myoglobin at 300 K. Since myoglobin consists of eight helices connected by turns (see Fig. 12), the minimized 10-ps coordinate sets were analyzed in terms of these secondary structural elements. The individual helices were very similar in all the structures; i.e., the rms differences for the α-carbon atoms of any given helix were mainly in the range from 0.4 to 0.8 Å, with the largest value (1.1 Å) occurring for helix E. However, when the positions of different helices that are in van der Waals contact were compared, the relative displacements were somewhat larger. The average translational displacements were between 0.8 and 1.4 Å and the average rotations between 1.6 and 5.2°. The various helix displacements were accomplished by rearrangements of the loop regions connecting them and by correlated motions of the sidechains which make the helix-helix contacts.

These results obtained from a molecular dynamics simulation of a single molecule can be compared with analysis of the X-ray average structures[231] of molecules that differ either because two molecules with the identical sequence are in inequivalent positions in a crystal, as in insulin,[232] or because the molecules from different species have highly homologous but not identical sequences, as in the globins.[233] In insulin,[232] the crystals contain hexamers with two different monomer environments; each monomer consists of 51 residues with three α-helices and one strand of β-sheet that is hydrogen bonded to an adjacent monomer. In comparing the two different monomers, it was found that the helices generally moved as rigid bodies (rms differences internally of 0.15 to 0.2 Å) and that the relative displacements of adjacent helices were 1.5 Å or less. For the globin series,[233] particularly when pairs with low sequence homology were compared (20 to 30%), the helix movements were somewhat larger than the average displacements found in the simulation;

i.e., for helices in contact, the relative displacements ranged up to 2.5 Å and the helix-helix angle varied between 8 and 10°. The hemoglobin structural differences, which correspond to average structures and are due to changes in the nature of the sidechains, are of the same order as the *maximum* differences between two minima found in the molecular dynamics of a *single* myoglobin molecule. This is in accord with the idea that the different molecules sample essentially the same region of conformational space and are governed by similar effective potentials between relatively rigid helices. Further, it suggests that the plasticity of a protein is sufficient to permit the substitution of different-sized amino acids in evolutionary development.

B. DOMAIN MOTIONS

Many enzymes[234-236] and other protein molecules (e.g., immunoglobulins) consist of two or more distinct domains connected by a few strands of polypeptide chain that may be viewed as "hinges." In lysozyme, for example, it was noted in a comparison of two X-ray structures[237] that the cleft appears to close down somewhat relative to the free enzyme when an active-site inhibitor is bound. The closing of the cleft resulted from relative displacements of the two globular domains that form the cleft. Other classes of enzymes (kinases, dehydrogenases, citrate synthase) have been found to have considerably larger displacements of two domains on substrate or inhibitor binding than does lysozyme.

In an analysis of lysozyme, which constituted the first theoretical approach to such hinge-bending motions,[152] the stiffness of the hinge was evaluated by the use of two different procedures. With an empirical energy function (see Chapt. III) an angle-bending potential was obtained by rigidly rotating one of the globular domains with respect to the other about an axis which passes through the hinge and calculating the changes in the protein conformational energy. This procedure overestimates the bending potential, since no allowance is made for the relaxation of the unfavorable contacts between atoms generated by the rotation. To account for the relaxation, an adiabatic potential was calculated by holding the bending angle fixed at various values and permitting the positions of atoms in the hinge and adjacent regions of the two globular domains to adjust themselves so as to minimize the total potential energy. As in the adiabatic ring-rotation calculation (see Chapt. VI.B.1), only small (< 0.3 Å) atomic displacements involving bond angle and local dihedral angle deformations occurred in the relaxation process. The frequencies associated with them (> 100 cm^{-1}) are much greater than the hinge-bending frequency (≈ 4.3 cm^{-1}; see below), so that the use of the adiabatic potential is appropriate. The hinge-bending potentials were found to be approximately parabolic, with the restoring force constant for the adiabatic potential about an order of magnitude smaller than that for the rigid potential (see Fig. 35). However, even in the adiabatic case, the effective force constant is about 20 times as large as the bond-angle bending force constant of an α carbon (i.e., $N—C_\alpha—C$); the dominant contributions to the force constant come from repulsive nonbonded interactions involving on the order of 50 contacts. If the adiabatic potential is used and the relative motion is treated as an angular harmonic oscillator composed of two rigid spheres with moments of inertia corresponding to those of the domains, a vibrational frequency of 4.3 cm^{-1} is obtained. This low frequency is a consequence of the fact that although the force constant is large, the moments of inertia of the two lobes are also large.

Although fluctuations in the interior of the protein, such as those considered in myoglobin (Chapt. VI.B.2), may be insensitive to the solvent (with the protein matrix acting as its own solvent), the domain motion in lysozyme involves two lobes that are surrounded by solvent. To take account of the solvent effect in the simplest way, the Langevin equation (Eq. 80) for a damped harmonic oscillator was used. The friction coefficient for the solvent damping term was evaluated by modeling the two globular domains as spheres.[238] From the adiabatic estimate of the hinge potential and the magnitude of the

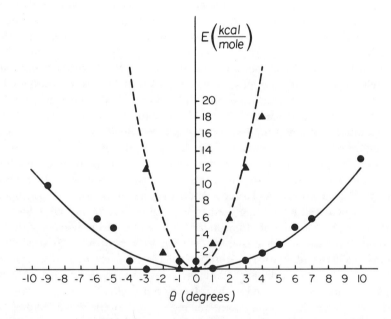

Figure 35. Change of conformational energy produced by opening ($\theta < 0$) and closing ($\theta > 0$) of the lysozyme cleft; calculated values are for the rigid bending potential (triangles) and for the adiabatic bending potential (circles); the origins for the two calculations are superposed.

solvent damping, it was found that the relative motion of the two domains in lysozyme is overdamped; i.e., in the absence of driving forces the domains would relax to their equilibrium positions without oscillating. Actually, the lysozyme molecule experiences a randomly fluctuating force from collisions with the solvent molecules, so that the distance between the globular domains fluctuates in a Brownian manner over a range limited by the bending potential; a typical fluctuation opens the binding cleft by 1 Å, corresponding to an angular motion of about 3°, and lasts for 20 ps. Thus, the solvent damping increases the "period" of a fluctuation from 7 ps for the vacuum system to 20 ps in solution.

A calculation[239] analogous to that for lysozyme by itself has been made for lysozyme with the inhibitor, tri-N-acetylglucosamine bound in the active-site cleft; a crystal structure for the complex was used for analysis. Since the adiabatic minimization procedure was improved somewhat from that used in the original lysozyme study just described,[152] both native and inhibited lysozyme were examined. The resulting frequencies were 3.0 cm^{-1} for native lysozyme (slightly lower than the value of 4.3 cm^{-1} obtained in the earlier work[152]) and 4.6 cm^{-1} for the molecule with bound inhibitor. The calculation with the inhibitor also suggested that the minimum energy structure was slightly more "closed" (about 10°) than that of native lysozyme.

Recently, the lysozyme hinge-bending mode has been examined by a more fundamental approach[239] than the adiabatic minimization technique. An iterative procedure was developed to determine the normal mode, or a set of normal modes, of the molecule that has the largest overlap with an initial model for the motion; in the lysozyme case, the initial model was the rigid-body motion used for the adiabatic calculations. The method is based on a modification of the Lanczos algorithm[240] that provides a general approach for finding a mode or modes related to a specific motion of interest; thus it is not necessary to solve the complete normal-mode problem for the molecule. The resulting value of the normal-mode frequency is 3.7 cm^{-1}, between the values of 3.0 and 4.3 cm^{-1} from the two adiabatic calculations described above. In a study of the normal modes of lysozyme in dihedral angle space,[136a] the lowest-frequency mode at 3.5 cm^{-1} appears to be of the hinge-bending type, although no evaluation of its overlap, or inner product, with the rigid rotation mode was given.

The hinge-bending normal mode[136] is considerably different in form from that obtained in the adiabatic calculations,[152,239] even though the overlap of the two is equal to 0.87. In the adiabatic minimization, the structural changes that were coupled to the hinge-bending motion were located in the hinge region. By contrast, the normal mode shows much more delocalized changes, as illustrated in a representation of the mode (see Fig. 36). This is also made clear by Fig. 37, which shows dihedral-angle changes associated with the

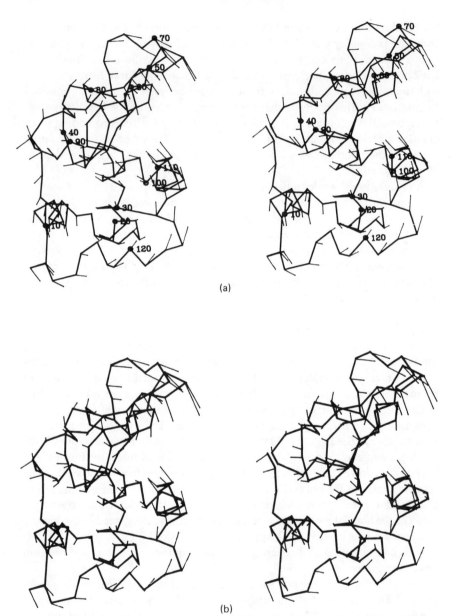

(a)

(b)

Figure 36. Stereo views of residue displacements in the lysozyme hinge bending. The thin line on each C$^\alpha$ shows the direction and magnitude of the mass-weighted rms residue displacements for a closing, 2-Å mass-weighted rms step along the normalized mode. (a) Rigid model; (b) normal mode that has the best overlap with the rigid model.

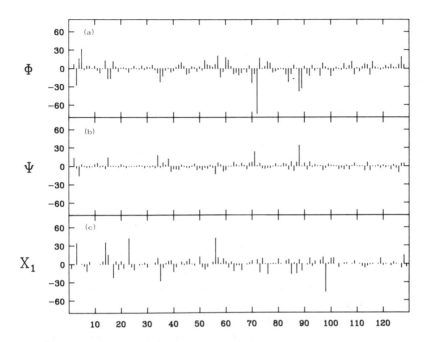

Figure 37. Dihedral-angle variations for hinge mode: (a) ϕ versus residue number; (b) ψ versus residue number; (c) sidechain angles χ^1 versus residue number. Each bar corresponds to the change in the dihedral angle in going from the 1-Å rms displaced closed structure to the 1-Å rms displaced open structure.

mode motion; in the rigid model only angles ϕ and ψ of residues 39 and 97, which are on the rotation axis, would change. That the energy associated with the mode is relatively insensitive to its detailed form is in accord with expectations; i.e., the *form* of the mode is linear with respect to the error of the calculation, while the *energy* has a quadratic dependence on the error.

Of particular interest is the fact that tryptophan residues play an essential role in the hinge-bending motion. In part, this is due to their large size and rigidity (see Fig. 3) and the presence of three of them (Trp-62, Trp-63, and Trp-108) in or near the active-site cleft. Trypthophans 62 and 108 are found to rotate in a concerted way in the hinge-bending mode, so as to become almost parallel as the hinge closes; Trp-28, which is in a hydrophobic region removed from the active site, is the most important residue that favors the open structure. These results suggest that tryptophans may have a general effect in stabilizing particular structures and modulating the transitions between them.

The adiabatic mapping technique used in the lysozyme study has now been applied to a number of other proteins; they include antibody molecules where the interdomain motions occur on a nanosecond time scale,[241,242] L-arabinose binding protein,[243] and liver alcohol dehydrogenase.[244] For the L-arabinose binding protein, calculations and experiments both suggest that the binding site is open in the unliganded protein and is induced to close by a hinge-bending motion upon ligation.[243,245] In the case of liver alcohol dehydrogenase, an open structure is stable in crystals of the apoenzyme[246] and a closed structure is stable for the holoenzyme.[247] The hinge-bending motion involves rotation of the catalytic domain relative to the coenzyme-binding domain. Adiabatic energy-minimization calculations[244] suggest that the apoenzyme is highly flexible. In fact, the adiabatic potential is such that normal thermal fluctuations could lead to a closed structure (a rotation of $10°$) similar to that found in holoenzyme.

Since the hinge-bending motion in lysozyme and in other enzymes involves the active-site cleft, it is likely to play a role in the enzymatic activity of these systems. In addition to possible differences in the binding equilibrium and solvent environment in the open and closed states, the motion itself could result in a coupling between the entrance and exit of the substrate and the opening and closing of the cleft.[248-250] In immunoglobins the interdomain mobility may be involved in adapting the structure to bind different macromolecular antigens and, more generally, it may play a role in the cross-linking and other interactions required for antibody function. In the coat protein of tomato bushy stunt virus, a two-domain structure, a hinge peptide has been identified from the X-ray structure,[251,252] and rotations about the hinge have been shown to be involved in establishing different subunit interactions for copies of the protein involved in the assembly of the complete viral protein shell.

Most of the information available on the hinge-bending or interdomain motions comes from high-resolution crystal structures. Low-angle X-ray scattering analyses of solutions provided evidence for radius of gyration changes that are in accord with the crystal results where available[253] or provide evidence for structural changes in cases where only one structure is known.[245,254] However, there is little experimental evidence on the time scales of the rigid-body motions. Fluorescence depolarization studies of labeled antibody molecules show that the time scale for internal motions ($\sim 10^{-8}$ s) is on the order of the calculated diffusional displacements of flexible hinged domains.[241,255] Similar results have been obtained for the myosin head group of muscle.[255a] Measurements[256] of the viscoelastic properties of triclinic lysozyme crystals have shown that the anisotropy of the compliance (the inverse of Young's modulus) is consistent with the molecular flexibility introduced by the hinge-bending mode. Studies of crystals with the inhibitor N-acetylglucosamine bound to the enzyme found a considerably decreased compliance, suggesting that the mode frequency has increased. Also, incoherent inelastic neutron scattering experiments (see Chapt. XI.D) show differences in the low-frequency region (<40 cm^{-1}) between native and inhibited lysozyme, but it is difficult to relate the results to specific structural changes. Recent neutron scattering experiments on lysozyme as a function of hydration have shown that there is increased mobility in the frequency range 1 to 5 cm^{-1} at higher hydration.[255b] Although Raman spectroscopy might appear to be well suited for studying the hinge-bending mode, extensions of the solvent damping calculations described above predict that the mode is significantly broadened and shifted toward zero frequency, making observation difficult.[257] There is a broad band at 25 cm^{-1} in the Raman spectrum of lysozyme crystals[258] that is absent in solution.[259] Whether this difference is due to solvent damping in the latter or the presence of a crystal mode at that frequency in the former is not clear.

C. SUBUNIT MOTIONS

There are many proteins that are composed of more than one subunit (i.e., independent polypeptide chains without covalent connections). They range from dimers made up of a pair of identical subunits to much more complicated systems that have many subunits which are not all identical; e.g., pyruvate dehydrogenase is composed of three types of subunits, with 24 of each in the entire complex.[260] In addition to such protein oligomers with a specific number of subunits, there are other systems (muscle, microtubules) that are protein polymers made up of an unspecified number of subunits. In globular proteins, the occurrence of two or more subunits is often involved with the regulation of activity. There may exist two quaternary structures, with one having high activity and the other, low activity. Because of the internal flexibility of proteins the relative motion of subunits (i.e., a change in quaternary structure) is required to transmit information over distances corresponding to the size of a subunit. In this way the binding of an effector to one subunit can alter the activity of another, as is the case in the well-studied protein aspartate transcarbamoylase of E. coli.[261,262]

The oligomeric protein for which the most detailed experimental and theoretical information is available is hemoglobin.[46,263,264] It has long been regarded as the prototypic system for the investigation of cooperativity in macromolecules. Hemoglobin is a tetramer composed of two types of subunits, two α chains and two β chains.[263] The two types of chains are very similar in sequence and tertiary structure; the tertiary structure is close to that of myoglobin (Fig. 12). Each subunit contains a heme with a ferrous iron at the center to which oxygen binds reversibly. As the fractional oxygen saturation of a tetramer increases from zero to 1, the affinity for oxygen increases; i.e., the binding is cooperative. The essential aim of studies of hemoglobin is to determine the detailed relationship between the structural changes induced by oxygen binding and the thermodynamic and dynamic manifestations of cooperativity. High-resolution X-ray studies have shown that there are two quarternary structures for the tetramer,[46] one of which (the deoxy structure) has low affinity for oxygen and the other (the oxy structure) has high affinity. The transition from the deoxy to the oxy structure can be described approximately as a rigid-body motion of two $\alpha\beta$ dimers, in which one dimer rotates by 15° with respect to the other and moves toward it by 0.8 Å.[265] The rate of the deoxy-to-oxy transition has been found to be on the microsecond time scale.[266] On a more detailed level, ligation of a subunit results in smaller but significant tertiary structural changes.[265,267] These mostly take place on a time scale of nanoseconds or less.[268] The coupling between the tertiary change induced by ligand binding and the quaternary transition is an essential element in understanding the cooperative mechanism. A statistical mechanical model of hemoglobin has been developed to provide a link between the observed structural changes and the thermodynamics of cooperative oxygen binding.[269,270]

To determine the nature of the motions involved in the cooperative transition, a reaction path for the tertiary structural change induced by ligand binding within a subunit was worked out by the use of empirical energy calculations.[267,271] It was shown that the motion of the iron into the heme plane on ligation introduces a perturbation of the heme that leads to the displacement of certain protein atoms, which in turn produce alterations in the surface regions that are in contact with other subunits. This provides a basis for the coupling between tertiary and quarternary structural change, in accord with available structural data.[265,272] The path followed in the quaternary transition has also been studied.[273] It has been shown that constraints introduced by the intersubunit contacts in the deoxy and oxy quaternary structures destabilize the liganded and unliganded subunits, respectively.

A molecular dynamics simulation[274] has been made of the photodissociation of carbon monoxide from an isolated α-subunit. This is of interest because optical absorption measurements[275] with 250-fs pulses showed that a deoxy-like species appears in 350 fs. It was suggested that the species formed has a high-spin ferrous heme with the iron out of the heme plane at or near the normal position found in unliganded hemoglobin. In the simulations, based on introducing a simplified potential for a high-spin pentacoordinate iron into the liganded α-chain structure, the time required for the iron to move out of the heme plane was between 50 and 150 fs. Similar times were obtained for an isolated heme group, suggesting that the local interactions with the protein do not significantly affect the iron motion relative to the heme that results from dissociation. To complete the description of the cooperative mechanism, full dynamical simulations of the tertiary and quaternary structural changes are required.

CHAPTER VIII

LARGER-SCALE MOTIONS

A variety of motions that are of interest involve significant changes in the secondary and tertiary structures of proteins. The polypeptide chain retains the covalent connections, but its conformation undergoes a large alteration. Some cases that have a known biological function are listed in Table I. The most widespread of these is the folding transition that occurs after a globular protein is synthesized and the subsequent unfolding when it is degraded. A number of proteins, such as calcium-binding proteins like calmodulin and paravalbumin,[276] seem to have a very flexible structure, particularly in the absence of their characteristic ligands, such as calcium. It has been suggested that this high internal mobility and the dependence of the structure on ligands makes them able to respond rapidly to their environment and thereby exercise certain control functions.[39,277] For some enzymes, a transition from a partly disordered structure to a fully ordered structure is involved in activation, as in the trypsinogen-trypsin system,[278] or in ligand binding, as in triosephosphate isomerase[279] and penicillopepsin.[280] Other systems where larger-scale structural changes are involved in the biological function include virus capsid RNA binding proteins,[251,252,281,282] DNA binding proteins,[283,284] and the influenza virus hemoglutinin protein that is active in cell fusion.[285] In the binding of the λ-phage repressor protein to the λ-phage operator DNA, for example, the N-terminal "arm," which consists of the amino acids Ser-Thr-Lys-Lys-Pro and is disordered in a crystal of the repressor, wraps around the DNA and significantly increases the binding constant.[283,284] For nucleic acids, also, there is evidence for larger scale motions.[286]

Helix-coil transitions, in particular, are of widespread occurrence in peptides and in proteins. One peptide hormone that has been studied experimentally is glucagon, which has a nearly random coil structure as a monomer in aqueous solution,[287] is partially helical when it interacts with a membrane[288] or trimerized in aqueous solution,[289] and is fully helical in a crystal where each molecule is involved in the formation of two trimers.[290]

In what follows we describe some of the larger-scale motions in proteins for which theoretical studies of the dynamics are available.'

A. HELIX-COIL TRANSITION

The helix-coil transition in peptides and proteins is one of the basic elements in the understanding of larger-scale motions. To make possible a dynamic simulation on the nanosecond-to-microsecond time scale required for the helix-coil transition in α-helices, a simplified model for the polypeptide chain was introduced. The model represents each residue by a single interaction center ("atom") located at the centroid of the corresponding sidechain, and the centers are linked by "virtual" bonds.[205] For the potential energy of interaction between the residues, assumed to be valine in the results described below, a set of energy parameters obtained by averaging over the sidechain orientations was used.[291] Terms that approximate solvation and the stabilization energy of helix formation were included. The diffusive motion of the chain "atoms" expected in water was simulated by using a stochastic dynamics algorithm based on Brownian dynamics (Chapt. IV.D). Starting from an all-helical conformation, the dynamics of several residues at the end of a 15-residue chain was monitored in several independent 12.5-ns simulations at 298 K. A typical trajectory is shown in Fig. 38. The mobility of the terminal residue was quite large, with an approximate rate constant of 10^9 s^{-1} for the transition between coil and helix states. This mobility decreased for residues farther into the chain. Unwinding of an interior residue required simultane-

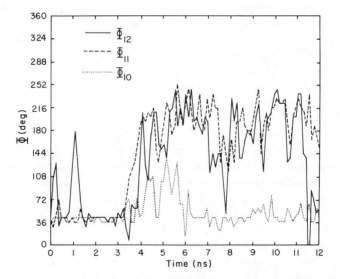

Figure 38. Dihedral-angle histories during a helix-unwinding trajectory: ϕ_{12} (solid line), ϕ_{11} (dashed line), and ϕ_{10} (dotted line).

ous displacements of several residues in the coil, so that larger solvent frictional forces are involved. The coil region did not move as a rigid body. Instead, the torsional motions of the chain were correlated so as to minimize dissipative effects. Such concerted transitions are not consistent with the conventional idea that successive transitions occur independently. Analysis of the chain diffusion tensor showed that the frequent occurrence of correlated transitions resulted from the relatively small frictional forces associated with such motions.[150] Further, the correlated nature of the torsional transition suggests that unwinding occurs in a relatively localized fashion and that a limiting value of about 10^7 s^{-1} will be reached for the interior of the helix. This value is in the experimentally observed range.[292]

B. PROTEIN FOLDING

The problem of the folding of a globular protein to its native structure has two parts.[293,294] The first is the static aspect concerned with the elements in the amino acid sequence that provide the structural information; the second deals with the dynamics of the folding process itself. Clearly, the answer to the first part of the problem is involved in the second (in particular, the stabilities of intermediate structures may be important in determining the folding paths) and conversely, it is possible, although less likely, that the second affects the first (i.e., that the nature of the folding process results in a nonequilibrium structure). Available data indicate that proteins in vitro can form their native structure in times from tenths of seconds to minutes in the absence of S—S bridges;[294] formation of S—S bridges coupled to refolding tends to take longer. This is to be contrasted with the much longer times that would be required to find the native conformation in a random search of all possible structures; in the simplest model, such a random search would take on the order of 10^{50} years for a protein consisting of 100 amino acids. Thus, protein folding must make use of more sophisticated search procedures which in one way or another divide a protein molecule into parts such that the information contained in the sequence of each part can be used independently. The simplest solution would be that a protein consisting of 100 amino acids is composed of 10 or so approximately equal parts, each of which can find its own unique stable form, and that these independently folded parts come together to form the native structure. From studies of peptides in solution,[288,289,295-300] as well as approximate energy calculations,[301] it appears that most sequences short enough to be searched through rapidly do not have a folded structure that is stable. Thus, alternatives to this simple idea must be examined. The diffusion-collision model[302-304] is one alternative mechanism that is described in the following paragraphs.

The diffusion-collision model considers the protein molecule to be divided into several parts (microdomains), each short enough for all conformations to be searched through rapidly. This condition implies that the native secondary structure of each such microdomain, although accessible by random events, is not stable. Consequently, several (two or more) of the microdomains have to diffuse together and collide in order to coalesce into a structural entity which is stable. The process of folding the entire protein to the native state then involves a series of such diffusion-collision steps. These might have to follow a unique order to yield the native structure (single correct pathway); alternatively and more likely, different sequences of diffusion-collision steps might be possible. Furthermore, particularly in larger proteins, there may be several regions (domains) that individually attain their native structure by a set of diffusion-collision steps and finally come together to form the native protein molecule; for certain proteins it has been shown that a separated domain can fold by itself.[305]

To illustrate the factors determining the kinetics of the diffusion-collision model, an example involving two microdomains is analyzed.[302,303] Each of the microdomains is considered to be in equilibrium between the native secondary structure and the unfolded random-coil structure, which includes the rest of the available conformational space. If the microdomains form helices in the native state (as in myoglobin), the helix random-coil transition described in this chapter (Sect. A) would be involved. It is assumed that coalescence can occur only if both partners have the native structure when they collide and that the secondary structural transitions are fast relative to other processes. If the two connected units are regarded as diffusing together from a finite distance to form a stabilized entity with the native conformation, the order of magnitude of the time τ required for coalescence can be obtained by considering the radial diffusion of spherical particles. The result is[303]

$$\tau \simeq \frac{1}{\beta} \frac{l \Delta V}{DA} \tag{84}$$

where ΔV is the volume of the finite diffusion space (the space between two concentric spherical shells), A is the target surface area (the area of the inner shell) whose radius is determined by the sizes of the peptide units, D is the diffusion coefficient, and l is a characteristic length (the average of the inner- and outer-shell radii, the latter being determined by the maximum distance of separation of the two units, which are part of the same protein chain). The quantity β accounts for the fact that only a fraction of the microdomains have the native secondary structure when they collide. In a detailed theory, the various parameters involved (i.e., ΔV, A, l) would have a direct physical significance (see below), but in this heuristic discussion they are phenomenological in character and only order-of-magnitude values can be given. With $D \sim$

1×10^{-6} cm^2/s, a value appropriate for spherical particles with the dimensions of protein microdomains, and with the outer-shell radius twice the inner radius and equal to 2×10^{-7} cm (20 Å), Eq. 84 yields $\tau \sim \beta^{-1}$ (10^{-7} to 10^{-8}) s. To evaluate β, an equilibrium, two-state model can be used for simplicity; i.e., $\beta = K_1 K_2/(1 + K_1)(1 + K_2)$, where K_1 and K_2 are the (native)/(random coil) equilibrium constants for the two separated units. This expression ignores the possibilities that β could vary as the partners approach and that details of the kinetics might modify the equilibrium result. From the formulas for τ and β above, it follows that for values of K_1 and K_2 equal to 10^{-2}, the folding time for a pair of units is of the order of 10^{-3} to 10^{-4} s. Stepwise folding by this mechanism seems most likely; for example, two units would coalesce to form a slightly more stable entity, which in turn collides with a third entity, and so on. In this way the rate of the process would increase as the folding advanced and a cooperative transition would result.

To examine the folding kinetics that might be expected for a protein, the diffusion-collision model has been applied to the operator-binding domain of the λ-repressor.[306] From the crystal structure,[283] the operator-binding domain consists of four helices that form a well-defined globule and a fifth that interacts primarily with an equivalent helix in the other subunit of a dimer. With the calculation restricted to the four helices of a monomer, the parameters for the coalescence reaction rate were estimated from Eq. 84 which was applied to the known structure; the possibility of dissociation of helices that had coalesced was included. Allowing only the correct helix contacts to form and solving the coupled rate equations with the estimated rate constants gives the results shown in Fig. 39. All the intermediates that have a maximum population of greater than 1% during the reaction are shown; 1 represents the denatured protein, 64 is the native protein, and the other numbers represent intermediates, with one or more of the helix contacts found in the native protein. There are 64 possible intermediates (e.g., 57 has the correct contacts for the pairs of helices 2-3, 2-4, and 3-4), but only a very small number contribute significantly. Although the exact details of the distributions depend on the choice of parameters, as does the absolute folding time, the general behavior is typical of what is expected from the diffusion-collision model.

To obtain a more realistic microscopic description of the coalescence of helical microdomains, stochastic dynamics simulations[151,307] have been performed on a model consisting of two stable helices, each composed of eight residues, with a random-coil segment of eight residues connecting them. The form of the potential was the same as that used for the helix-coil transition calculation[206] (this chapter, Sect. A). The stochastic dynamics simulation gave results rather close to those of the simple model. However, this appears to be due to the cancellation of two effects neglected in that model. The first is that the intervening chain *increases* the coalescence rate because it favors

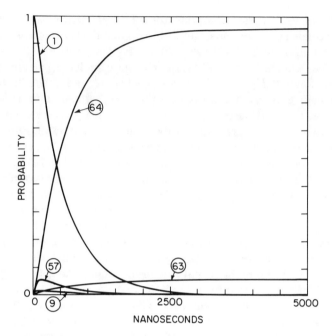

Figure 39. Probabilities of significantly populated intermediate states for the folding of the λ-repressor in the diffusion collision model as a function of time for $\beta = 0.01$. The states are labeled as described in the text.

configurations that are close together (essentially an entropic term related to the number of chain configurations) and the second is that hydrodynamic interaction between the helices decreases the effective mutual diffusion constant and, therefore, *lowers* the coalescence rate.

Clearly, these studies of protein folding dynamics are only the first steps in obtaining the solution to a very complex problem. Nevertheless, they provide insights that are relevant to experiments and suggest refinements needed in improved treatments of protein folding.

C.   DISORDER-TO-ORDER TRANSITIONS

The occurrence of disorder-to-order transitions involving segments of a protein, the remainder of which has a well-defined conformation, has been documented in crystal structures of a variety of systems. The segments involved may be at the N terminal or the C terminal end (disordered "arm") or in the middle of the polypeptide chain (disordered "loop"). The length of the disordered segment varies from five or six amino acids to a sizable portion of the protein. The transition is generally characterized by the fact that in one crystal structure the polypeptide chain segment is not visible, while in another it is observed to have a well-ordered structure. The fact that a segment is not observed in an X-ray structure may be due to its being a random coil with a wide range of contributing conformations or to its having only three or four different conformations in the molecules making up the crystal; usually, a small amount of disorder is sufficient to make it impossible to obtain definitive X-ray results. Thus, additional experimental data and theoretical studies are needed to determine the nature of the disorder that actually occurs. In a few cases, this has been obtained from NMR measurements (e.g., for the N-terminal arm of the λ repressor, and virus coat proteins)[284,308] that show that the segment involved is as flexible as a random coil, and in one case from perturbed angular correlation measurements that give the time scale of the disordered loop motions (the trypsinogen-trypsin transition), as described below.[309]

1. Trypsinogen-Trypsin Transition

The enzyme trypsin exists as a catalytically inactive proenzyme, the zymogen trypsinogen, and the active form, trypsin, which is obtained by cleavage of an N-terminal hexapeptide from the former. X-ray crystallographic studies have shown that 85% of the structures of the two molecules are identical.[278] The remaining 15%, called the activation domain, consists of three external loop segments 7 to 12 residues in length that are partly or entirely disordered in trypsinogen crystals but ordered in trypsin. It has been suggested that trypsinogen is inactive because the ordered activation domain is required for forming the substrate binding site; i.e., although trypsinogen could exist in the ordered form needed for activity, the excess entropy loss required to undergo the disorder-to-order transition inhibits substrate binding.

Experimental results have suggested that the two residues at the N-terminus in trypsin (Ile and Val) are essential for stabilizing the ordered structure; the N terminus of Ile forms a salt link with Asp-194 and both Ile and Val interact with the activation domain at the active site. Discharge, removal, or blocking of the Ile in the homologous chymotrypsin-chymotrypsinogen enzyme-zymogen system leads to inactivation.[310] Also, perturbed angular correlation measurements with mercury labels[309] show that there exists a dynamical process in the 10-ns range for trypsinogen that is absent in trypsin. It is likely that the motion involved is that of the disordered activation domain, which has considerably more conformational freedom in trypsinogen than in trypsin.

Full molecular dynamics simulations of more than 1 ns are still almost prohibitive for large systems, so that a direct calculation of the relaxation time is difficult. However, it is possible to examine the flexibility of the activation domain as a first step toward analyzing the trypsinogen-trypsin transition. Starting with the known structure of trypsin, a hypothetical trypsin-to-trypsinogen transition has been studied by a series of molecular dynamics simulations.[311] The difference between trypsin and trypsinogen was mimicked by removing the key N-terminal peptide (Ile-Val) in the former; this is simpler than moving the disordered N terminus out into solution, where it appears to be located in trypsinogen. Simulations were done of the normal and trypsinogen-like trypsin, in both cases starting with the ordered trypsin structure, and the differences in structure and dynamics introduced by the perturbation were determined. A stochastic boundary simulation approach of the type described in Chapt. IV.C was employed. Since the activation domain includes a sizable portion of the protein, artifacts due to constraining protein atoms in the buffer region had to be avoided. A reaction region of 22-Å radius that included about 90% of the trypsin molecule was used. In addition, 668 water molecules were included to solvate the important portions of the trypsin molecule. Although this results in a large simulation, it is still much smaller than that for trypsin plus about 5000 water molecules with periodic boundary conditions.[312]

As a first step in the study, the trypsin model was simulated for 10 ps after equilibration. This simulation was found to be in good agreement with the X-ray results for the average atomic positions and the fluctuations relative to the average positions. Starting with the structure obtained at the end of the 10 ps, residues Ile-16 and Val-17 were deleted, five waters were introduced into the binding pocket to replace the deleted residues, and the modified system was simulated for 10 ps; as a control the unperturbed trypsin system was also simulated for another 10 ps. Comparisons were made of the deviations of the 10-ps simulations for the unperturbed and perturbed structures from the original 10-ps unperturbed simulation. It was found that removing the N-terminal peptide induced a conformational change of the activation domain; i.e., the average structure of the activation domain in the perturbed simulation differed significantly from the unperturbed simulation, while that in the control did not. The residues involved (i.e., the residues of the activation domain) are essentially identical to those found to be disordered in trypsinogen. Furthermore, the glycine residues that delimit the activation domain loops undergo the largest changes in backbone dihedral angles.

These results suggest that the N-terminal dipeptide is the trigger and that the glycine residues are the hinges for the trypsinogen-to-trypsin transition. Further work will be necessary to follow the dynamics of the transition and to estimate the entropic change involved in going from a disordered to an ordered structure.

2. Triosephosphate Isomerase

Another type of disorder-to-order transition is the so-called "loop-cap" transition in certain enzymes.[279,280] This structural change is associated with a highly mobile loop of residues, from 6 to 12 amino acid residues in length, that are located near the active site. Such a structural feature has been identified in the enzyme triose phosphate isomerase (TIM). TIM makes an especially interesting case study for theoretical analysis of the loop-cap transition because there exist X-ray,[279,313] kinetic,[121] and thermodynamic data[121] for this enzyme.

TIM is a dimeric protein with each subunit having a molecular weight near 27,000 D. It catalyzes the interconversion of dihydroxyacetone-phosphate and glyceraldehyde-3-phosphate in the glycolytic cycle, which plays a critical role in the ATP production of aerobic organisms.[313] The structure of a TIM monomer is shown in Fig. 1c. Each monomer consists of a core of eight strands of parallel β-sheet forming a cylinder with a slight right-handed twist. This core is surrounded by α-helices connected with segments that form well-defined turns and longer loops. One loop of 10 residues, numbered 168 to 177 in the isolated yeast enzyme, is disordered in the X-ray structure, even at a resolution of 1.9 Å.[314] When the crystals were exposed to a solution containing the substrate dihydroxyacetone phosphate at $-5°C$, a different X-ray structure resulted. The disordered region was clearly resolved and the substrate was bound in the active site with the ordered loop covering the active site. The structural data suggest that some of the residues may have moved as much as 10 Å upon substrate binding.

The phosphate group of the substrate in the X-ray structure interacts with several residues at the active site, as well as with residues 171 to 173 in the loop region. Thus attention has focused on the role that the structure and fluctuations of the loop region may play with regards to the specificity and activity of the enzyme. Clearly, the folding of the loop over the substrate excludes bulk solvent from the active site; this is likely to be important in preventing hydrolysis reactions that could compete with the isomerization. Also, the loop is thought to restrict the nature of possible substrates to molecules that can bind to the active site and anchor the loop in the closed conformation. One of the steps associated with the release of product appears to be rate limiting[121] in the enzyme-catalyzed reaction; it is possible that this step is the loop opening.[314] Dynamic simulations of the-transition of the loop and its relation to enzyme function are in progress.[314a]

It is likely that the entropy loss in closing the loop partially offsets the enthalpic gain on binding. In this way the enzyme can be highly specific, yet not bind the substrate too strongly. Further, it is possible that the disorder-to-order transition in the formation of the enzyme substrate complex raises its free energy relative to that of the transition state. In this way the fluctuations of the loop could play a catalytic role in the classical sense of reducing the free-energy barriers between reactants and products.[314]

An interesting speculation suggests an additional role for fluctuations in TIM in the process of catalysis.[314] This involves so-called "directed fluctuations" and is related to a variety of suggestions that have been made for the channeling of displacements and energy in proteins.[28,22] The protein as a whole and the surrounding solvent are assumed to provide energy in the form of thermal fluctuations that is used in the chemical reaction catalyzed by the enzyme. If this occurred in a purely random way, the behavior of the enzyme would be no different from any other heat bath. However, since the protein is an ordered nonrandom structure, it is possible that certain fluctuations involve motions along the reaction coordinate. This could lead to an enhancement of the reaction rate. If the loop region were made up of individual, uncorrelated Brownian particles, as it would be in the disordered state, the motions would be essentially random. The other extreme corresponds to a well-defined structure with regular, deterministic motions due to one or more normal modes. The idea that specific low-frequency modes of an enzyme contribute to catalysis is esthetically appealing, even if unlikely. For pyruvate kinase, an enzyme with a rather slow rate, NMR experiments have been interpreted as indicating that no such special motional effects contribute to the catalysis.[315] However, the interpretation of the experiments is not unequivocal and the results for one system do not exclude the possibility for other enzymes. Clearly, more experimental studies, as well as specific simulations, are needed to elucidate this question.

CHAPTER IX

SOLVENT INFLUENCE ON PROTEIN DYNAMICS

The solvent, usually an aqueous phase, has a fundamental influence on the structure, thermodynamics, and dynamics of proteins at both a global and a local level.[316,317] The significance of the solvent has been demonstrated in a variety of biologically important phenomena, ranging from the rate of oxygen uptake in myoglobin[318] to the stabilization of oppositely charged sidechain pairs at the surface of proteins.[319] The simulation methods described in Chapt. IV.B and C can provide information concerning the details of the interactions between proteins and the solvent. In this chapter we review several applications of simulation methods that focus on the influence of the solvent, and more generally the environment, on the structure and the dynamics of peptides and proteins. The results discussed refer to a variety of simulation methods and include a range of solvent influences on biopolymers from global effects on protein structure to the influence of solvent on the atomic fluctuations.

A. GLOBAL INFLUENCES ON THE STRUCTURE AND MOTIONAL AMPLITUDES

To illustrate the solvent effect on the average structure of a protein, we describe results obtained from conventional molecular dynamics simulations with periodic boundary conditions.[92,193] This method is well suited for a study of the global features of the structure for which other approaches, such as stochastic boundary simulation methods, would not be appropriate. We consider the bovine pancreatic trypsin inhibitor (BPTI) in solution and in a crystalline environment. A simulation was carried out for a period of 25 ps in the presence of a bath of about 2500 van der Waals particles with a radius and well depth corresponding to that of the oxygen atom in ST2 water.[193] The crystal simulation made use of a static crystal environment arising from the surrounding protein molecules in the absence of solvent. These studies, which were the first application of simulation methods to determine the effect of the environment on a protein, used simplified representations of the surroundings because of limitations on the available computer time. It was expected that certain essential properties of any solvent (namely, its attractive interactions with the protein atoms and the excluded volume effects due to its presence) would be represented satisfactorily by the van der Waals solvent. Further, comparisons with simulations using more realistic water models, when they became available, would make possible a separation of the effects due to directional hydrogen bonds and specific water structure from more general solvation properties. In Tables VI.A and B are listed results for the average radius of gyration of BPTI, and the atomic number density and density fluctuations for spherical shells centered at the protein center of mass. For the radius of gyration, the van der Waals solvent and crystal environments have a similar influence on the protein. They yield a radius of gyration close to the value calculated from the crystal structure; by contrast, the radius of gyration from a 25-ps simulation in vacuo with the same potential function is 8% smaller.[148] In more recent vacuum simulations[192,320,321] such shrinkage or "compaction" of the protein has been prevented by the use of larger van der Waals radii or by including all hydrogen atoms in the empirical energy function. Adjusting the van der Waals radii to mimic the effect of solvent is some-

TABLE VIA
Radius of Gyration (in Å) for BPTI

Vacuum	10.2
Crystal	10.89
Solution	10.63
X-ray	10.96

Source: Ref. 193.

TABLE VIB
Number Density for BPTI

$r_{c_2} - r_{c_1}$ (Å)^a	$\langle \rho \rangle$ (Å^{-3})		$\langle (\Delta \rho)^2 \rangle^{1/2}$ (Å^{-3})	
	Crystal	Solution	Crystal	Solution
3–0	0.0827	0.0582	0.0089	0.0100
6–3	0.0606	0.0682	0.0028	0.0035
9–6	0.0599	0.0588	0.0019	0.0020
12–9	0.0320	0.0325	0.0010	0.0012
15–12	0.0115	0.0131	0.0005	0.0007

^{a}The density and density fluctuations are computed for atoms in spherical shells $r_{c_2} - r_{c_1}$, where r_{c_i} corresponds to a distance from the center of the molecule.

Source: Ref. 193.

what artificial, although it can be regarded as a simplified potential of mean force correction. When the density profiles are compared, the net effect of the van der Waals solvent is to decrease the density in the interior of the protein (up to 9.0 Å from the center) and to increase the density slightly throughout the rest of the molecule (9.0 to 15.0 Å from the center), relative to the vacuum simulation. Thus, the overall force field due to the van der Waals solvent is attractive, somewhat more so than that from the fixed crystalline environment. These calculations demonstrated that solvent and environmental effects on proteins can be probed by simulation methods and that many of the observed phenomena can be rationalized in terms of basic features of the protein-solvent interactions, such as the global "attractive" force field provided by the van der Waals solvent.

With the availability of faster computers, BPTI was simulated in aqueous solution and in a solvated crystal with a more realistic (three-center) water model.[92] The simulations were limited to 8 ps of equilibration and 12 ps of analysis, somewhat short for definitive conclusions to be drawn; recently, a crystal simulation of BPTI that extended over 40 ps has been reported.[322] The average structures obtained from the various simulations are compared in Table VII. In the three calculations made with the same empirical potential, the van der Waals solvent and static crystal field results yielded an average structure closer to the experimental crystal structure than did the vacuum calculation. The full crystal simulations, including crystal waters, gave an average structure still closer to the X-ray result, while the deviation from the crystal structure of the average structure obtained from the aqueous solution simulation was similar to the earlier vacuum result.

One difficulty with the above comparisons is that different empirical en-

TABLE VII

RMS Deviations of BPTI Simulations from X-Ray Structure (Å)

Simulation Type	All Atoms	C^α Atoms
Vacuum[a]	3.02 (1.5)[b]	2.20 (1.01)[b]
Static crystal[a]	2.12	1.52
Full crystal[c]		
$\langle 8-20 \rangle$	1.62	1.06
$\langle 19-20 \rangle$	1.95	1.33
Van der Waals solvent[a]	1.94	1.35
Explicit water solvent[d] $\langle 15-20 \rangle$	2.69	1.81

[a] Ref. 193.

[b] Values in parentheses from a vacuum simulation of 50 ps with a more recent (CHARMM 19) potential function (H. Yu and M. Karplus, unpublished results).

[c] Ref. 322. Values in brackets indicate the time periods over which the results were averaged.

[d] Ref. 92. Values in brackets indicate the time period over which the results were averaged.

ergy functions were used in the two sets of simulations. The effect of the empirical potential is illustrated by the fact that a recent 25-ps vacuum simulation of BPTI that used a revised polar hydrogen representation yielded an average structure with an all-atom rms deviation from the X-ray results of only 1.5 Å, a value smaller than that obtained in any of the simulations listed in Table VII.[323]

It is of interest to compare in more detail the average atomic positions obtained from the fully solvated crystal and aqueous solution simulations.[92] Figure 40 shows the magnitude of the difference between the time-averaged positions of protein atoms in solution and in the crystal; the C^α atom results are plotted in the lower portion and those for the end atoms of sidechains are shown in the upper portion. Since the time average is limited to a 12-ps period, a substantial error is likely to be associated with this comparison; e.g., C^α atom differences less than 1.5 Å and sidechain atom differences less than 2.0 Å are not considered significant. With this criterion, the only significant mainchain differences occur for C^α atoms involved in crystal contacts and external loops, i.e., residues 24 to 29, residues 12 to 15, and the carboxy-terminal residues (see Fig. 17). For sidechain atoms, crystal contacts account

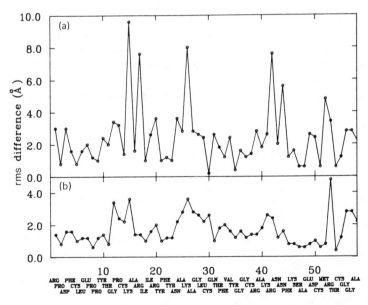

Figure 40. Solvent effect on atomic positions. The rms difference (Å) between the average structures from 12.0-ps simulations of BPTI in solution and in a hydrated crystal are plotted versus residue number: (a) the difference for terminal sidechain atoms and (b) for mainchain C^α atoms. (From Ref. 92.)

for some differences; e.g., residue Arg-17 is in contact with Ala-27 and Ala-58 of a neighboring molecule. Polar and charged sidechains show different conformations in the crystal and in solution, in accord with the results of earlier energy minimization studies of BPTI crystals.[209]

The influence of the environment on the magnitude of the fluctuations of the protein atoms is illustrated in Table VIII, which shows the atom-averaged rms fluctuations for BPTI in vacuo, in the van der Waals solvent, and in the static crystal environment;[193] the results are computed from averages over 25-ps simulations for each case. When the overall trend is considered (i.e., averages over all atoms), the relative ordering for the magnitudes of the fluctuations is van der Waals solvent > vacuum > crystal. Thus, there is a "softening" of the effective potential in going from vacuum to solution and from the crystal to solution; that the crystal environment appears to be more rigid than vacuum is likely to be a result of the artificial character of the model used for the crystal. The full crystal simulation including solvent gave rms fluctuations of the C^α atoms equal to 0.62 Å; this is similar to the value (0.52 to 0.58 Å) found for corresponding time periods in the earlier simulations. The largest increase in the magnitude of the fluctuations in solution, relative to the vacuum result, comes from exposed atoms (particularly C^γ, C^δ, and N^ζ). A corresponding increase would be expected for exposed atoms which are not involved in crystal contacts, but this is not found in the simulation. The presence of solvent does not affect the magnitude of fluctuations of interior sidechain atoms and most mainchain atoms. Corresponding solvent effects in an aqueous environment are discussed below.

In summary, the molecular dynamics simulations show that the net effect of solvent on the overall structure of a protein is rather small and that the structure in the crystal is similar to that obtained in solution. This point is of some relevance since X-ray studies of the crystalline state are the basis for much of our current understanding of the structure and function of proteins.

TABLE VIII
Effects of Solvent and Crystalline Environment on the Root-Mean-Square Fluctuations, $\langle \Delta r^2 \rangle^{1/2}$ in Å, of atoms in BPTI[a]

Atom type	Vac	Sol	Crys	Atom type	Vac	Sol	Crys
All	0.72	0.78	0.67	C^β	0.62	0.69	0.62
N	0.52	0.50	0.50	C^γ	0.74	0.89	0.71
C	0.53	0.51	0.50	C^δ	0.78	1.15	0.77
O	0.66	0.65	0.67	N^ζ	0.88	1.40	1.00
C^α	0.54	0.54	0.52				

[a] Vac, Sol, and Crys refer to vacuum, solvent, and crystal simulations, respectively.
Source: Ref. 193.

However, it must be cautioned that the available simulations are far too short to demonstrate that the average structures obtained from them are equilibrium structures;[92] i.e., to overcome the barriers along the paths from the neighborhood of the native structure to some different or denatured state could require microseconds or longer. A case in point is the similarity of the vacuum, van der Waals solvent, and aqueous solution results. It is likely, at least from the viewpoint of conventional wisdom, that in the vacuum and van der Waals environment, the native structure is metastable.

Solvent causes a net "softening" of the local atomic potentials, particularly for exposed sidechain atoms, with a commensurate increase in fluctuation amplitudes. Recent inelastic neutron scattering studies of lysozyme as a function of hydration are in accord with this conclusion.[255b] In addition, there are significant localized conformational changes, particularly for sidechains and exposed loop regions, and these, coupled with the increased fluctuations observed in the solvent simulation, may be of functional importance. These points are amplified in Chapt. X, where a thermodynamic analysis of the effect of solvent on the conformational equilibria of a dipeptide model is given.

B. INFLUENCE ON DYNAMICS

The presence of solvent can affect the dynamics of proteins in two ways. The first is due to alteration of the equilibrium properties of the molecule; i.e., changes in the average structure and the potential of mean force governing the atomic fluctuations can be induced by solvent. In addition, there are purely dynamical effects which result from frictional forces (nonzero viscosity) and random collisional impulses that act on the atoms of the protein. Solvent influences on the structural and equilibrium fluctuations of biopolymers can be dramatic; e.g., the change in conformer populations of the alanine dipeptide produced by an aqueous environment relative to vacuum (see Chapt. X.A) or the stabilization of like-charged ion pairs at the surface of lysozyme (this chapter, Sect. D). The effect of solvation on the magnitude of the atomic fluctuations was described above. Solvent can also alter the time evolution of the atomic motions, and consequently, observable properties dependent on these motions. Examples are the viscosity dependence of carbon monoxide rebinding to myoglobin,[318] the variation of fluorescence anisotropy decay with the solvent environment,[324,325] and the rate of enzyme catalysis, including the binding of substrate and the release of product in different solvent environments.[103,326] In this section we describe the results of simulation studies that provide insights into the manner in which solvent alters the internal motions of peptides and proteins.

1. Alanine Dipeptide Results

One effect of solvent on the fluctuations of biomolecules is to "dampen" or slow down their time evolution.[152,327-329] This simple picture, based on the ideas of Brownian motion first proposed by Debye, Einstein, Smoluchowski, and others,[212] attributes the solvent influence to a Stokes-like dissipative force, $4\pi\eta v_0 a$, where η is the solvent viscosity, v_0 is the particle velocity, and a is the radius of the particle. Thus, the solvent viscosity is anticipated to have a significant influence on the time development of the motions. Molecular dynamics simulations with a detailed solvent model can be used to examine such effects.[163] Here we illustrate the influence of viscosity and the solvent-modified potential of mean force by describing some results from stochastic dynamics simulations of the alanine dipeptide, N-methylalanyl acetamide.[330] The simulations were performed by numerical integration of the Langevin equation (Eq. 16) for two different potential surfaces; one corresponded to the dipeptide in vacuum and the other to a theoretical potential of mean force surface for the dipeptide in aqueous solution (see Chapt. X.A). Further, for the aqueous potential of mean force, two simulations with different values for the solvent viscosity were compared; one used the viscosity of water at 300°K ($\eta \simeq 1$ cP) and the other a somewhat higher value. As is evident from Eq. 16 and the expression for the Stokes dissipative force given above, the solvent viscosity appears in both the dissipative and fluctuating force terms; in vacuum with zero viscosity the Langevin equation of motion reduces to Newton's equation. All of the simulations consisted of approximately 60 ps of dynamics carried out at 300 K in the neighborhood of the C_{7eq} configuration ($\phi = -60°$, $\psi = 60°$), which is a local minimum on both the vacuum and the solvated potential surfaces. In Fig. 41 the spectral density, $I(\omega)$, of the dihedral-angle fluctuation autocorrelation function is plotted; $I(\omega)$ is defined by

$$I(\omega) = \int_0^\infty \cos \omega t \, \langle \Delta\psi(t) \, \Delta\psi(0) \rangle \, dt \qquad (85)$$

where $\Delta\psi$ is the dihedral-angle fluctuation relative to its mean value; the spectral densities shown in the figure are normalized to their maximum values.

The vacuum potential results, corresponding to the limit of zero viscosity, are shown in Fig. 41a. At zero viscosity, the dihedral angle ψ oscillates with a period of approximately 0.63 ps. When the conditions are changed to represent water at 300 K (i.e., the solvent-modified potential-of-mean-force surface is used and $\eta = 1.0$ cP), the dominant effect is that the dihedral motion has a periodicity of about 3.7 ps (see Fig. 41b). The solvent influence observed in these simulations is consistent with an earlier molecular dynamics study of

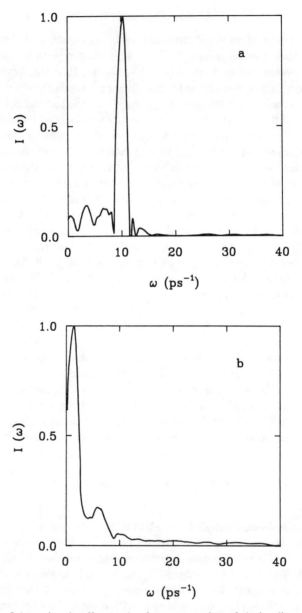

Figure 41. Solvent viscosity effects on low-frequency motions of alanine dipeptide. The normalized spectral density for the ϕ dihedral angle is plotted versus frequency (ps^{-1}) for (a) dynamics on a vacuum potential surface (see Fig. 58a); (b) dynamics with a potential of mean force (see Fig. 58b) in a solvent of viscosity, $\eta = 1.0$ cP; (c) dynamics with a potential of mean force (see Fig. 58b) in a solvent of viscosity, $\eta > 1.0$ cP.

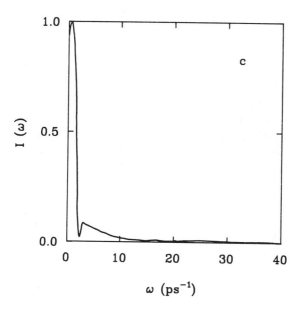

the alanine dipeptide that treated the aqueous solvent explicitly and used periodic boundary conditions.[85,163]

The influence of viscosity can be separated from that due to a change in the potential of mean force by considering the third simulation, which used the same solvent-modified potential but a higher viscosity. Figure 41c shows that there is an additional increase in the period of oscillation to ~4.2 ps, as indicated by the peak in the spectral density at $\omega \simeq 1.5$ ps^{-1}. This makes clear that the dynamical influence of solvation on the internal motions of the alanine dipeptide is to slow their time evolution. In an analysis of the normal modes on the vacuum and solvent-modified potential it is found that the shifts of the vibrational frequencies and the changes in the form of the modes are small.[115] Further, the root-mean-square fluctuation of the dihedral angle ψ, which depends only on the potential of mean force and is not affected by solvent viscosity, is very similar on the two surfaces. The rms fluctuations are 12.3° and 13.8° for the vacuum and aqueous environment simulations, respectively.[115,330] This confirms that the changes in the effective period of the oscillations of ψ described above are due predominantly to the solvent viscosity; only a slight "softening" is found in the potential of mean force, as indicated by the increase of 1.5° in the fluctuation amplitude. However, it should be noted that for the rate of barrier crossing between the accessible minima, both the viscosity and the overall flattening of the potential surface due to solvent are expected to be important (see Chapt. X.A).

The influence of solvent viscosity on the dynamics of small biopolymers, like the alanine dipeptide, is clearly illustrated by the results given here. However, it is necessary to determine whether for larger biopolymers like globular proteins, the solvent influence on the dynamics can also be related to the viscosity alone. Below we provide evidence to the contrary. It is found that the solvent does not affect all atoms and all of their dynamical properties in the same way. Thus, a description based on only the solvent viscosity is not adequate, even disregarding possible alterations in the potential of mean force.

2. Protein Results

The simulations of BPTI in vacuum and in a van der Waals solvent, described above (this chapter, Sect. A), were analyzed to determine the effect of solvation on the time scales of the atomic motions.[199] Given the displacement autocorrelation function, $C(t)$, for a fluctuation, the relaxation time is defined by[211]

$$\tau = \int_0^\infty C(t)\,dt \tag{86}$$

For a correlation function that can be represented approximately as an exponentially decaying function of time, τ can be obtained from a linear fit to $\log[C(t)]$ versus t. Distributions of the relaxation times calculated in this way for BPTI are shown in Fig. 42. Histograms of the number of atoms as a function of relaxation time for all atoms in the vacuum and solvent simulations are given in Fig. 42b and a, respectively; separate plots for the backbone (N, C^α, C) and sidechain atoms from the solvent simulation are presented in Fig. 42c and d, respectively. Most of the atoms have relaxation times in the range between 0.4 and 2.5 ps, with the distribution skewed toward longer times. From a comparison of Fig. 42a and b it can be seen that the relaxation times in the solvent are shifted to slightly longer times relative to the vacuum results. This is in general agreement with predictions based simply on the viscosity of the environment. When the solvent simulation relaxation times are separated into mainchain and sidechain contributions (Fig. 42c and d), it is evident that the relaxation times most strongly influenced by the solvent correspond to sidechain atoms. This effect was quantified by comparing the mean and second moment of the distributions displayed in Fig. 42c and d with the comparable distributions from the vacuum simulation. The values obtained are $\tau_{vac}^m = 1.65 \pm 1.26$ ps for mainchain atoms and $\tau_{vac}^s = 1.42 \pm 0.98$ ps for sidechain atoms in the vacuum simulation compared with $\tau_{sol}^m = 1.79 \pm 1.05$ ps for mainchain atoms and $\tau_{sol}^s = 2.04 \pm 1.69$ ps for sidechain atoms in

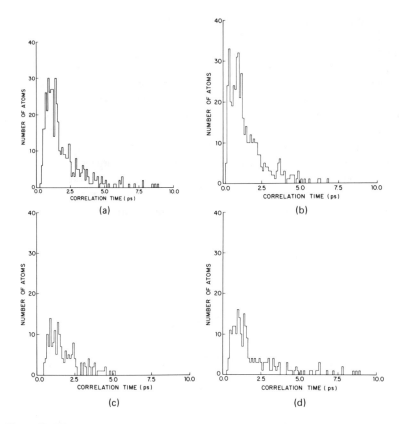

Figure 42. Histogram of relaxation time (τ) distributions showing the number of atoms with a given value of τ as a function of τ: (a) all atoms, solvent run; (b) all atoms, vacuum run; (c) backbone atoms, solvent run; (d) sidechain atoms, solvent run.

the solvent simulation; i.e., very little change occurs in the mainchain atom relaxation times, while a large change is observed for sidechain atoms. In small proteins such as BPTI, a significant fraction of sidechain atoms are exposed to solvent, while most of the mainchain atoms are not. Thus it is clear that viscosity cannot influence all atoms uniformly. Instead, a gradation of effects appears to occur, with the degree of solvent damping depending on the interaction between the protein atoms and the solvent; the latter can be described approximately by the solvent exposure of the protein atoms.

To explore these effects more thoroughly, results are presented from stochastic boundary molecular dynamics simulations of the active-site cleft of lysozyme in the presence of aqueous solvent and in vacuum.[108] The simulation treated the atoms in a 11.0-Å region around Trp-62 (see Fig. 8*b*); in both the vacuum and the solvent simulation, 294 protein atoms were included, with 132 water molecules present in the solvent simulation. Four simulations were carried out, each consisting of 10 to 15 ps of thermalization/equilibration and 40 ps of dynamics. Two solvated and two vacuum dynamics simulations were performed with corresponding but slightly different initial conditions to provide an estimate of the statistical error of the calculations. Figure 43 shows the displacement autocorrelation functions of the atom $N^{\epsilon 1}$ of Trp-62 and of the atom C^{β} of Asn-46. The solid lines represent the correlation functions from an average over the 80 ps of dynamics in the presence of solvent and the dashed lines are the 80 ps vacuum results. The most striking feature is the absence of solvent influence on the evolution of the correlation function of Asn-46 C^{β}. This is to be contrasted with the large damping of the correlation

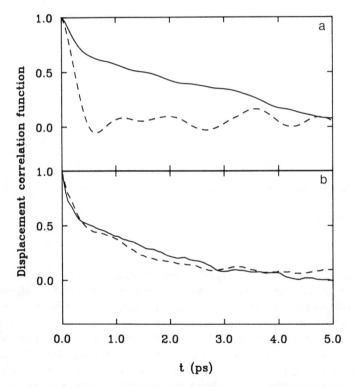

Figure 43. Solvent effects on local protein motions. The normalized displacement autocorrelation functions are plotted versus time for residues near the active site in lysozyme. Vacuum simulation results are plotted as dashed lines and solvent simulation results are plotted as solid lines for (*a*) Trp-62 $N^{\epsilon 1}$ and (*b*) Asn-46 C^{β}.

function for Trp-62 $N^{\epsilon 1}$. Insight into the origin of this difference is provided by comparing the fractional accessibility of the two residues; the fractional solvent accessibility is defined as the surface area of the residue in the protein as measured by the Lee and Richards algorithm for a solvent probe of 1.6 Å radius,[331] divided by the exposed surface area for the same residue with the same configuration in the absence of surrounding protein. The calculated accessibilities are 0.3 for Trp-62 and 0.08 for Asn-46. Thus it appears that the dynamical influence of the solvent can be related to the degree of solvent exposure, as was concluded from the survey of relaxation times in BPTI described above. This leads to the conclusion that the effect of solvent on the dynamics of local displacements in the interior of the protein is small. As has already been pointed out, larger-scale collective motions (e.g., those involved in a hinge-bending mode) are expected to be affected by the solvent viscosity because a significant fraction of the atoms involved in the motion are exposed to solvent. The observed solvent viscosity dependence of ligand rebinding in myoglobin[318] suggests that dynamics of surface residues is involved in the motion of the ligand through the protein.

In Fig. 44 the velocity autocorrelation functions are shown for the same atoms as in Fig. 43. The solvent has a small effect on the evolution of this dynamical property for both the exposed residue (Trp-62) and the buried residue (Asn-46), in contrast to the behavior found for the displacement autocorrelation functions. To interpret this difference, we note that the displacements evolve on a time scale much slower than velocities; this is made clear by the fact that the time scales used in Figs. 43 and 44 differ by a factor of about 10. Thus the solvent influence is linked to the time scale of the protein motions relative to those of the solvent, as well as to the exposure of the atoms involved. To elucidate this point, the spectral density of the atomic displacement and velocity autocorrelation function for Trp-62 $N^{\epsilon 1}$ and for the velocity autocorrelation function of ST2 water are displayed in Fig. 45. The spectral density, $C(\omega)$, for a property is defined as the cosine transform of the autocorrelation function $C(t)$, as in Eq. 85. From Fig. 45 it is clear that the motional time scale for water ranges from 0 ps^{-1} to near 150 ps^{-1}. The dominant contributions to the velocity autocorrelation functions for the protein atom are near 250 ps^{-1}, and those for the displacement autocorrelation functions are in the range 0 to 30 ps^{-1}. Thus there is a strong overlap of motional time scales between protein and solvent for the displacement correlation function, while there is essentially no overlap for the velocity correlation function. Because the ST2 model uses rigid water molecules, the internal bond length and bond angle vibrations are not present. These occur at frequencies near 300 ps^{-1} and higher frequencies, and although there is some overlap, they would not be expected to contribute significantly to the relaxation of the protein atom velocity autocorrelation functions.

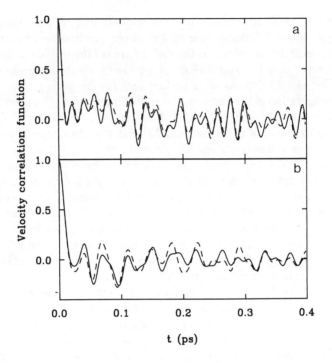

Figure 44. Same as Fig. 43 except for velocity autocorrelation function.

The difference in the spectral density between the displacement and velocity autocorrelation functions can be understood from a normal-mode description (see Chapt. IV.F). Using Eq. 23 for the displacement autocorrelation function and differentiating it to obtain the velocity autocorrelation function, one finds that the terms in the latter are weighted by the square of the mode frequency relative to the former. Thus higher-frequency contributions are more important in the spectral density associated with the velocity autocorrelation function than the displacement autocorrelation function.[153,332]

As a final example of the influence of solvent on dynamical processes in biopolymers, we examine the decay of the fluorescence emission anisotropy for Trp-62 and Trp-63 in lysozyme; the model used is described in Chapt. XI.C).[324] In this model the fluorescence anisotropy, $r(t)$, as a function of time for a given residue is given by (see Eq.110)

$$r(t) = \frac{2}{5}\, e^{-t/\tau_0}\, \langle P_2[\hat{\mu}_A(0) \cdot \hat{\mu}_E(t)] \rangle \tag{87}$$

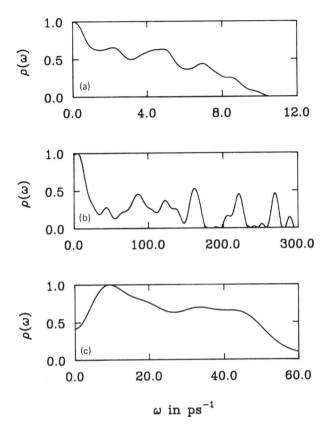

Figure 45. Time-scale matching for protein-solvent motions. The normalized spectral density for the (a) displacement and (b) the velocity autocorrelation functions of Trp-62 $N^{\epsilon 1}$, and (c) for the velocity autocorrelation function of ST2 water. (Note the differences in the timescales.)

The correlation function, $\langle P_2[\hat{\mu}_A(0) \cdot \hat{\mu}_E(t)] \rangle$, provides a measure of the internal motions of particular residues in the protein.[324,333] Figure 46 shows the results obtained for Trp-62 and Trp-63 from the stochastic boundary molecular dynamics simulations of lysozyme used to analyze the displacement and velocity autocorrelation functions. The net influence of the solvent for both Trp-62 and Trp-63 is to cause a slower decay in the anisotropy than occurs in vacuum. In vacuum, the anisotropy decays to a plateau value of 0.36 to 0.37 (relative to the initial value of 0.4) for both residues within a picosecond. In solution there is an initial rapid decay, corresponding to that found in vacuum, followed by a slower decay (without reaching a plateau value) that continues beyond the period (10 ps) over which the correlation function is ex-

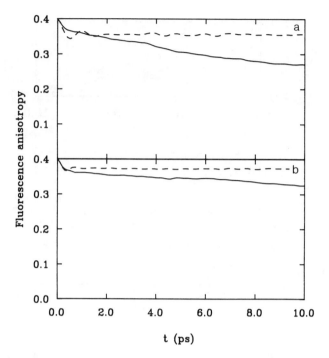

Figure 46. Solvent influence on fluorescence anisotropy for tryptophans in lysozyme. Vacuum simulation results are plotted as dashed lines and solvent simulation results are plotted as solid lines for (a) Trp-62 and (b) Trp-63.

pected to be reliable.[334] There is a difference in the solvent effect for the two residues, which appears to be related to the different spatial coupling with the solvent; the fractional accessibility for Trp-62 is 0.3 and that of Trp-63 is 0.01. The greater mobility of Trp-62 in solution, relative to the vacuum simulation, arises from the presence of mobile "bulk" solvent which can hydrogen bond to $N^{\epsilon 1}$ of the tryphophan. In the vacuum simulation a "structural" water was included[192] that formed hydrogen bonds with $N^{\epsilon 1}$ of Trp-62 and another protein residue, thereby restricting the tryphophan motion.

The simulation results presented in this section illustrate two essential features of the solvent and of solvent-protein interactions which influence protein dynamics. One is concerned with the spatial coupling (i.e., the degree of coupling between solvent and protein, which is related to solvent accessibility or some other measure of direct relatively short-range solvent-protein interactions), and the second is concerned with the time-scale coupling (i.e., the degree to which the motions of the solvent are commensurate with the temporal evolution of a dynamical property of the protein). Temporal and spatial coupling are known to be important generally in dynamical processes in solutions.[332] It is not surprising, therefore, that they play an essential role in protein-solvent interactions. In addition, for molecules as complex as proteins, very specific structural effects involving the solvent can alter the dynamics.

Simulation methodologies allow one to investigate a variety of solvent effects. For small, highly solvated biopolymers, stochastic dynamics with a potential of mean force provides a technique for studying the effect of solvent on low-frequency motions and conformational transitions (see this chapter, Sect. B.3, below). The validity of this approach to the dynamics is based on the approximately constant frequency response of the solvent in the low-frequency regime. Thus it is expected to be the most accurate for the study of relatively slow processes (i.e., for processes on the time scale of 100 ps or longer). For situations in which protein motions on varying time scales may be important (e.g., during the course of an enzyme-catalyzed reaction), the stochastic boundary molecular dynamics approach or conventional molecular dynamics techniques are necessary. If the processes of interest are localized, the stochastic boundary approach is appropriate because it provides the necessary detail with the least computational effort. Finally, for infrequent events occurring in complicated systems that involve the solvent, methods such as activated dynamics with a stochastic boundary can be employed.

3. Stochastic Dynamics Simulations of Barrier Crossing in Solution

Activated processes in solution, such as conformational transitions for biomolecules that are fully exposed to solvent, can be treated by stochastic dynamic simulation methods (see Chapt. IV.D). Their use requires a knowledge of the solvent contribution to the potential of mean force. Also, the system must be small enough so that the simulation times can be extended to the nanosecond or microsecond range required to adequately sample the statistically rare events involved in activated processes. Alternatively, activated dynamics techniques can be used with a stochastic dynamics simulation.[307,335]

A case of biological interest that could be treated by stochastic dynamics occurs in carboxypepdidase A[336] where the structural changes that occur on binding the substrate include a large displacement of a surface sidechain, Tyr-248, which moves more than 10 Å toward the active site. Although it was originally thought that the tyrosine participates directly in the enzymatic reaction (i.e., by its hydroxyl group binding to the catalytically active zinc), site-specific mutagenesis showed that the tyrosine could be replaced by a phenylalanine while maintaining activity.[337] This suggests that its role may be to protect the active site, analogous to the loop in TIM considered in Chapt. VIII.C.2.

As a model for such surface sidechain transitions, we consider here the motion of linear aliphatic chains $(CH_2)_n$—CH_3 that are attached to the surface of a protein and are fully exposed to solvent.[338] In the model calculations, the end of the chain attached to the macromolecule was held fixed and the Langevin equations of motion (Eq. 16) for the atoms of the chain were solved simultaneously for times up to a microsecond. The methyl and methylene groups of the chain were treated as single extended atoms with a friction coefficient corresponding to methane in water. A generalized empirical potential energy function was used to represent the intramolecular interactions (nonbonded and torsional) in the usual way, except that they were slightly modified to take into account the presence of solvent; that is, a potential of a mean force surface calculated by integral equation methods replaced the isolated molecule potential function (see Chapt. V.C);[339] the torsional barrier in the potential function is equal to about 2.8 kcal/mol. The simulations showed the motion with respect to a given torsional angle separates into two time scales. The shorter time motions, on the order of tenths of picoseconds, correspond to torsional oscillations within a potential well, and the longer time motions, on the order of 200 ps, correspond to transitions from one potential well to another. The transition rates found from the simulation agree well with a Kramers model for the activated process (Eqs. 82 and 83). To test the validity of this type of calculation, comparisons of the stochastic dynamics results with NMR relaxation measurements (e.g., ^{13}C NMR) have been made (see Chapt. XI.B).

C. SOLVENT DYNAMICS AND STRUCTURE

For an understanding of protein-solvent interactions it is necessary to explore the modifications of the dynamics and structure of the surrounding water induced by the presence of the biopolymer. The theoretical methods best suited for this purpose are conventional molecular dynamics with periodic boundary conditions and stochastic boundary molecular dynamics techniques, both of which treat the solvent explicitly (Chapt. IV.B and C). We focus on the results of simulations concerned with the dynamics and structure of water in the vicinity of a protein both on a global level (i.e., averages over all "solvation" sites) and on a local level (i.e., the solvent dynamics and structure in the neighborhood of specific protein atoms). The methods of analysis are analogous to those commonly employed in the determination of the structure and dynamics of water around small solute molecules.[163] In particular, we make use of the conditional protein "solute"-water radial distribution function,

$$g_{ps}(r) = \frac{d \langle N_{ps}(r) \rangle}{4\pi \langle \rho_{ps} \rangle r^2 \, dr} \tag{88}$$

where $N_{ps}(r)$ is the number of solvent molecules s (oxygen atoms for water) at a distance between r and $r + dr$ from protein atom p, and $\langle \rho_{ps} \rangle$ is the average solvent density within a spherical volume centered on protein atom p; it is convenient to define a local solvent density in the normalization of $g_{ps}(r)$ for cases involving globular proteins because shielding by other protein residues can reduce the density substantially below that in the bulk. The function $g_{ps}(r)$ details the structure of the solvent around particular protein atomic sites and is useful for interpreting solvent dynamical behavior. The integrated value of $g_{ps}(r)$, namely $\langle N_{ps}(r) \rangle = \int_0^r g_{ps}(s)4\pi s^2 \, ds$, corresponds to the coordination number for the solvent around site p at a distance r.

The dynamics of the solvent in the region near a protein can be characterized by a number of properties (e.g., solvent velocity correlation functions, mean-square displacement correlation functions, dipole orientation correlation functions, etc.). These properties provide information on a range of phenomena from local solute-solvent interactions (velocity correlation functions) to solvent mobility (mean-square displacement correlation functions) and dielectric behavior (dipole correlation functions). Here we focus on the diffusion constant, which provides a convenient measure of mobility for water molecules near protein atoms. The diffusion constant for solvent molecules may be computed directly from the slope of the mean-square displacement correlation function,

$$\lim_{t \to \infty} \frac{d}{dt} \langle |r(t) - r(0)|^2 \rangle = 6D \tag{89}$$

Trypsin in aqueous solution has been studied by a simulation with the conventional periodic boundary molecular dynamics method and an NVT ensemble.[312,340] A total of 4785 water molecules were included to obtain a solvation shell four to five water molecules thick in the periodic box; the analysis period was 20 ps after an equilibration period of 20 ps at 285 K. The diffusion coefficient for the water, averaged over all molecules, was 3.8×10^{-5} cm^2/s. This value is essentially the same as that for pure water simulated with the same SPC model,[341] 3.6×10^{-5} cm^2/s at 300 K. However, the solvent mobility was found to be strongly dependent on the distance from the protein. This is illustrated in Fig. 47, where the mean diffusion coefficient is plotted versus the distance of water molecules from the closest protein atom in the starting configuration; the diffusion coefficient at the protein surface is less than half that of the bulk result. The earlier simulations of BPTI in a van der Waals solvent showed similar, though less dramatic behavior;[193] i.e., the solvent "molecules" in the first and second solvation layers had diffusion coefficients equal to 74% and 90% of the bulk value. A corresponding reduction in solvent mobility is observed for water surrounding small biopolymers.[163] Thus it

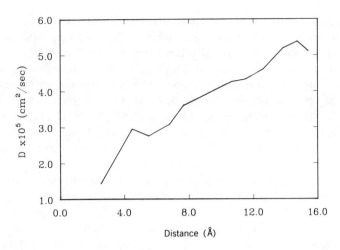

Figure 47. Diffusion coefficient for SPC water in the vicinity of a protein. The diffusion coefficient is plotted versus the distance (in Å) of the water molecule center of mass from the surface of trypsin. (From Ref. 340.)

is likely that specific interactions and the modifications in the water structure provide the explanation of the reduced mobility. Water molecules in the range 10 to 15 Å from the protein were found to have diffusion coefficients slightly higher than the bulk value. This increase was attributed to the presence of a relatively disordered region of solvent resulting from competition between the structural influence of the protein and its solvation shells on one hand, and the forces that determine bulk water structure on the other.

A more detailed description of solvent structure and dynamics around specific protein atoms has been obtained from the stochastic dynamics simulation of the active-site cleft region of lysozyme with an ST2 water model (this chapter, Sect. B.2 above);[108] the differences resulting from the use of the more structured ST2 model versus the less structured SPC model are expected to be small. The radial distribution functions and coordination numbers representing the water oxygen-protein atom solvation structure are displayed in Figs. 48 to 50. These properties are plotted for specific apolar (hydrophobic) atoms in Fig. 48, for specific polar atoms in Fig. 49, and specific charged atoms in Fig. 50. In addition, Table IX lists the local average solvent density $\langle \rho_{ps} \rangle$, within a 6-Å sphere and the position of the first peak in $g_{ps}(r)$ for each protein atom site. The most important difference between the solvation structure around apolar atoms (Fig. 48) and polar or charged atoms (Figs. 49 and 50) is the shape of the radial distribution function. The distribution of solvent around the apolar atoms, which peaks near 3.45 Å and is

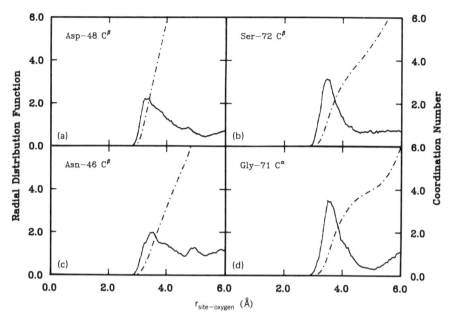

Figure 48. Solute-solvent radial distribution functions and running coordination numbers. The radial distribution (the solid line using the left scale) and the running coordination number (the dashed line using the right scale) are plotted versus distance in angstroms for the distribution of water oxygens around apolar atoms: (a) Asp-48 C^β; (b) Ser-72 C^β; (c) Asn-46 C^β; and (d) Gly-71 C^α.

broad, contrasts with the distribution function for Asp-52 $O^{\delta 1}$, which has a sharp peak at $r = 2.7$ Å. Such behavior is consistent with previous studies of small solute molecules[112,113,115,115a,163,341-344] and confirms that water near the apolar groups is structured differently from bulk water or water near polar or charged groups.[345] The generally broad distributions around C^β of Asp-45 and Asn-46 appear to be different from those around Ser-72 C^β and Gly-71 C^α, which are similar. This difference may be attributed to the fact that the protein "environment" around C^β of Asp and Asn is different because of side-chain intramolecular correlations that are absent for Gly-71 C^α and Ser-72 C^β. There results a less structured distribution function and a more uniform increase in coordination number for Asp and Asn; the coordination number varies almost linearly with r for these atoms. The difference is similar to that observed in the water structure around CH_2 and CH_3 groups of butane in aqueous solution.[343,344] Thus we see that intramolecular correlations can modulate the solvation structure around apolar atoms.

The solvation structures around the polar atoms Trp-62 $N^{\epsilon 1}$ (Fig. 49a) and

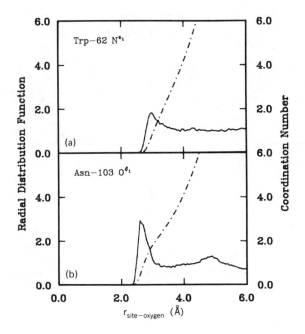

Figure 49. Same as Fig. 48 except for polar atoms: (*a*) Trp-62 N$^{\epsilon 1}$ and (*b*) Asn-103 O$^{\delta 1}$.

Asn-103 O$^{\gamma 1}$ (Fig. 49*b*) are similar to each other and to those observed in an earlier simulation of the alanine dipeptide.[163,342,346] There are on the average four water molecules within 4 Å of these atoms. The distance at which the first peak occurs and the number of solvent molecules within the first solvation sphere are typical for the solvation of polar species and indicate a hydrogen-bonding solvation structure similar to that of pure water.[343] Because the polar group interactions with water are much stronger than those involving nonpolar groups, modulation of the distributions due to neighboring atoms is less likely for the former than for the latter. The difference in peak heights between the two curves is a result of the large size of the indole ring of Trp-62, which blocks a significant fraction of the solid angle around N$^{\epsilon 1}$.

For the charged atoms (Fig. 50), all of the distribution functions show features typical of charged-group solvation.[112,113] There are four to five solvent molecules within the first solvation sphere, defined as extending to the first minimum in $g(r)$ that is at ~3.5 Å. This is indicative of tightly bound solvent around the charged group.

Results for the dynamics of water around protein sidechains in lysozyme

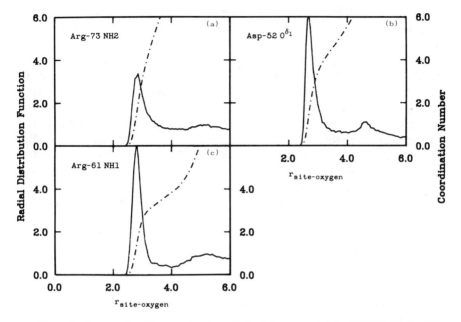

Figure 50. Same as Fig. 48 except for charged sidechain atoms: (a) Arg-73 NH2; (b) Asp-52 $O^{\delta 1}$; (c) Arg-61 NH1.

TABLE IX
Solvent Characteristics Near Protein Atoms

Atom	Solvent Density[a] (molecules/Å³)	First Peak Position in $g(r)$ (Å)	Diffusion Constant[a] (cm²/s) × 10⁵
Asp-48 C^β	0.02111	3.33	0.60
Asn-46 C^β	0.01294	3.45	0.52
Ser-72 C^β	0.00901	3.52	0.38
Gly-71 C^α	0.00764	3.50	0.52
Trp-62 N^ϵ	0.01781	2.95	0.77
Asn-103 O^δ	0.01673	2.67	0.73
Arg-73 NH2	0.02617	2.78	0.63
Arg-61 NH1	0.01323	2.80	0.51
Asp-52 O^δ	0.01762	2.70	0.56
Bulk water[b]	0.0315	—	1.30

[a] Water molecules within 6.0 Å from the atomic site are considered.
[b] Bulk water density taken as average from simulation.

Source: Ref. 108.

are displayed in Figs. 51 to 53. General features of the change in the mobility of water in the vicinity of the various atoms are indicated by the difference in the diffusion constants (see Figs. 51 to 53 and Table IX), which show the trend $D_{apolar} < D_{charged} < D_{polar} < D_{bulk}$. This behavior is consistent with results for the alanine dipeptide in water[163,346] and with those from the trypsin simulation described above.[312,340] The reduced mobility of water around apolar groups is consistent with the formation of clathrate-like solvation shells that disrupt the tetrahedral structure of bulk water. The resulting reduction in hydrogen-bond-forming potential inhibits the usual mechanism of translational diffusion.[163] A low diffusion coefficient for waters bound to charged sidechains is consistent with the stronger effective "solute" solvent interactions, which constrict the solvation sphere of waters and create a more rigid cage than for the waters around polar groups. This can be seen by comparing the structure and dynamics of waters around the charged atom Asp-52 $O^{\delta 1}$ (Figs. 50 and 53 and Table IX) with that around the polar atom Asn-103 $O^{\delta 1}$ (Figs. 49 and 52 and Table IX). The reduction in the mobility of water bound

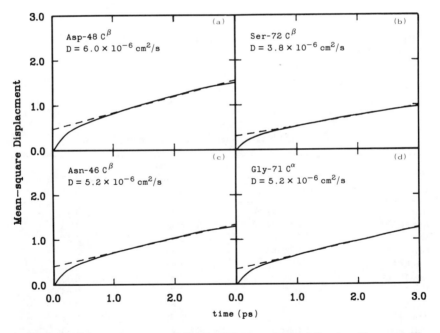

Figure 51. Solvent mean-square displacement for water in the cleft region of lysozyme. The mean-square displacement is plotted versus time (ps) for water within a 6.0-Å sphere around the apolar atoms: (a) Asp-48 C^{β}; (b) Ser-72 C^{β}; (c) Asn-46 C^{β}; (d) Gly-71 C^{α}. The dashed line indicates the linearly extrapolated diffusional motion.

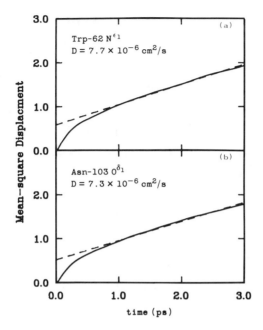

Figure 52. Same as Fig. 51 except for polar atoms: (a) Trp-62 N$^{\epsilon 1}$ and (b) Asn-103 O$^{\delta 1}$.

to polar groups in the lysozyme simulation is greater than that found for the alanine dipeptide.[163] This suggests that neighboring group interactions may be involved.

To summarize, the effect of a protein on the dynamics of the surrounding water is to produce an overall reduction in mobility in regions close to the protein. When the origins of this reduction are examined by analyzing the structure and dynamics of solvent around particular groups of atoms within the protein, a consistent picture emerges that connects solvation structure and dynamics. The dominant reduction in mobility involves water bound to apolar and charged groups; the water interacting with polar groups is more similar to bulk water.

D. ROLE OF WATER IN ENZYME ACTIVE SITES

Interactions between specific solvent molecules and protein atoms are important for protein structure and function. Water molecules participate directly in many enzymatic reactions, including those catalyzed by the large class of enzymes that hydrolyze peptide bonds. X-ray crystallography has identified waters that appear to be an integral part of the structure of a protein. Some of

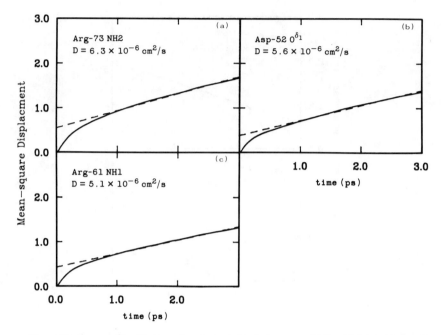

Figure 53. Same as Fig. 51 except for charged sidechain atoms: (a) Arg-73 NH2; (b) Asp-52 $O^{\delta 1}$; (c) Arg-61 NH1.

these are interior waters, i.e., water molecules that are completely surrounded by protein atoms and separated from bulk water (e.g., the four water
molecules in BPTI[347] and a larger number of waters in trypsin[348]). Other
structural waters are in contact with the bulk solvent but can be identified
because they occupy the same positions in different crystal forms of the same
protein or related proteins, as in several lysozymes that have been studied at
high resolution.[349] In most cases, such structural waters form at least two hydrogen bonds to protein atoms. It is often true, however, that important water
molecules cannot be identified by crystallography because they do not have
well-defined positions. Such is the case when the water molecules interact
with mobile sidechains (e.g., lysines) that are themselves disordered in the
crystal.[350]

    To characterize structural waters and to follow the dynamics of reactions
involving waters, it is necessary to be able to treat in detail the motions of
these molecules. A methodology which includes solvents explicitly is required. Both conventional molecular dynamics techniques and the stochastic
boundary molecular dynamics approaches can be used. When the region of

interest is localized, or energy exchange is expected to be important, stochastic boundary methods are more appropriate (Chapt. IV.C).

Bovine pancreatic ribonuclease A has been used for theoretical studies of the structural and dynamic role of active-site waters, as well as for a more general examination of enzyme catalysis by simulation methods. Although little is known about the physiological role of ribonuclease, a particularly wide range of biochemical, physical, and crystallographic data are available for this enzyme.[351,352] The accumulated data have led to proposals for the catalysis of the hydrolysis of RNA by a two-step mechanism[351-353] in which a cyclic phosphate intermediate is formed and subsequently hydrolyzed. Both steps are thought to involve in-line displacement at the phosphorus and to be catalyzed by the concerted action of a generalized acid and a generalized base. However, a full understanding of the mechanism and a detailed analysis of the rate enhancement produced by ribonuclease has not been achieved. To supplement the experimental data, stochastic boundary molecular dynamics simulations[103] have been carried out of the solvated active site of ribonuclease A by itself in the absence of substrates or other anions (sulfate, phosphate) bound to the active site, of ribonuclease A with the bound dinucleotide substrate CpA, and of ribonuclease A with the bound transition-state analogue uridine vanadate (UV).

The individual simulations and their comparisons provide insights into aspects of the active-site region of ribonuclease that are important for an understanding of substrate binding and catalysis. Figure 54 shows stereo pictures of some of the average dynamics structures. Figure 54a shows the active-site residues for ribonuclease A, Fig. 54b in the presence of CpA, and Fig. 54c in the presence of UV. Residues that have been implicated in substrate binding (Thr-45, Ser-123 in the "pyrimidine" site and Gln-69, Asn-71, Glu-111 in the "purine" site) and in catalysis (His-12, His-119, Lys-7, Lys-41, Lys-66, Phe-120, Asp-121, Gln-11) are shown, with some additional residues that may be involved in maintaining the structural integrity of the active site; in Fig. 54a the average positions of some of the water molecules are included. A comparison of these figures and analysis of the simulation results indicates that in the free enzyme, water molecules make hydrogen bonds to protein polar groups that become involved in ligand binding. A particularly clear example is provided by the adenine-binding site in the CpA simulation (Fig. 54b). The NH_2 group of adenine acts as a donor, making hydrogen bonds to the carbonyl of Asn-67, and the ring N1 acts as an acceptor for hydrogen bonds. These hydrogen bonds are present in the free ribonuclease simulation, with appropriately bound water molecules replacing the substrate. Thus these waters, and those that interact with the pyrimidine site residues Thr-45 and Ser-123, help to preserve the protein structure in an optimal arrangement for substrate binding. Corresponding substrate "mimicry" has been observed in X-ray struc-

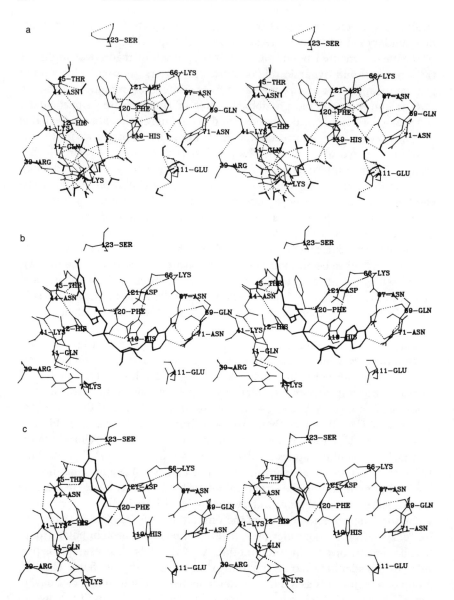

Figure 54. (*a*) Stereo drawing of native ribonuclease A, including active-site residues and bound-water molecules. The structure corresponds to the molecular dynamics average structure of the protein and water molecules. Protein-protein hydrogen bonds are indicated when the heavy donor-acceptor distance is less than 3.5 Å and the donor-hydrogen-acceptor angle is less than 65.0°. (*b*) Similar to (*a*) for the ribonuclease-CpA complex, including substrate and active-site residues. (*c*) Similar to (*a*) for the ribonuclease-UV complex, including substrate and active-site residues. No water molecules are shown in (*b*) and (*c*).

tures of lysozyme[349] and penicillopepsin.[280] The presence of such waters in the active site has the consequence that water extrusion and its inverse must be intimately correlated with substrate binding and product release; e.g., there are 23 fewer waters in the active site of the CpA complex than for free ribonuclease in the present simulation.

A striking feature of the active site of ribonuclease is the presence of a large number of positively charged groups, some of which may be involved in guiding and/or binding the substrate.[354] The simulation of the native enzyme demonstrated that these residues are stabilized in the absence of ligands by well-defined water networks. This is clear in Fig. 55, which shows an example that includes Lys-7, Lys-41, Lys-66, Arg-39, and the doubly protonated His-119, as well as the water molecules within 3 Å of any charged site with their energetically most likely hydrogen bonds. The bridging waters, some of which are organized into trigonal bipyramidal structures, are able to stabilize the otherwise very unfavorable configuration of positive groups because the interaction energy between water and the charged $C-NH_n^{\delta+}$ ($n = 1, 2, 3$) moieties is very large; e.g., at a donor-acceptor distance of 2.8 Å, the $C-NH_3^{\delta+} \cdots$ H_2O energy is -19 kcal/mol with the empirical potential used for the simulation,[65] in approximate agreement with accurate quantum-mechanical calculations[355] and gas-phase ion-molecule data.[356] During the simulation, the water molecules involved in the charged group interactions oscillate around their average positions, generally without performing exchange. On a longer time scale, it is expected that the waters would exchange and that the sidechains would undergo large-scale displacements. This is in accord with the disorder found in the X-ray results for lysine and arginine residues (e.g., Lys-41, Arg-39; see Refs. 350 and 357), a fact which makes difficult a crystallo-

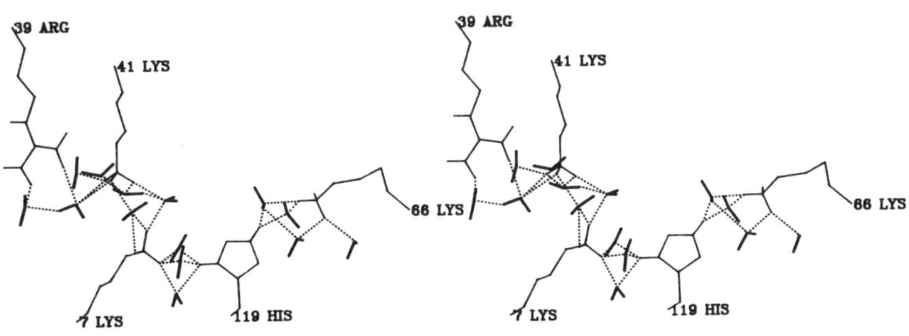

Figure 55. Stabilization of positively charged groups by bridging water molecules in the active site of native ribonuclease A without anions. The stereo drawing corresponds to a snapshot of the molecular dynamics trajectory at 15 ps. The picture includes only residues Lys-7, Arg-39, Lys-41, Lys-66, His-119, and the bound water molecules. The hydrogen-bonding criterion is the same as used in Fig. 54.

graphic determination of the water structure in this case. Water stabilization of charged groups has been observed[358,359] or suggested[360] in other proteins. It is also of interest that Lys-7 and Lys-41 have an average separation of only 4 Å in the simulation, less than that found in the X-ray structure. The fact that this like-charged pair can exist in such a configuration is corroborated by experiments which have shown that the two lysines can be cross-linked;[361] the structure of this compound has been reported recently[362] and is similar to the native protein.

In the active-site simulations of lysozyme[108] (this chapter, Sect. B.2 above) similar water networks that stabilize charged groups have been observed. To illustrate the dynamics of the formation of such networks, a sequence of stereo plots showing the formation and evolution of a stable pair of positively charged residues is displayed in Fig. 56. The pair consists of $(NH_2)^+$ moieties of Arg-61 and Arg-73. The solvated structure evolved from a conformation obtained in a vacuum simulation of lysozyme.[108,192] The sequence of plots shows the formation of the water-bridged pair over a time period from $t = 0$ ps to $t \approx 8$ ps, which followed dynamical equilibration of the solvent around the fixed vacuum structure of the protein. After 8 ps, the ion-pair structure is stable, but fluctuations in the pattern of hydrogen bonds do occur; typical

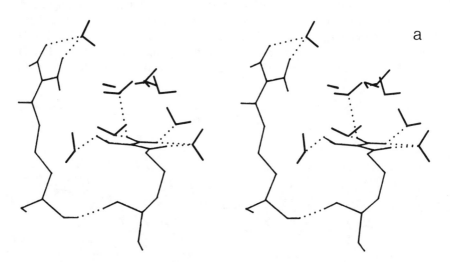

Figure 56. Formation of like-charged "ion" pairs illustrated by the solvation of two arginine sidechains in lysozyme (Arg-61, leftmost residue, and Arg-73). Stereo plots show (a) the "unsolvated" structure at $t = 0.0$ ps; (b) the initial formation of the water-bridged structure at $t = 8.25$ ps; (c) and (d) the fully "solvated" water-bridged structures for $t = 16.5$ ps and $t = 33.0$ ps, respectively. The protein-solvent hydrogen bonds are shown as dotted lines.

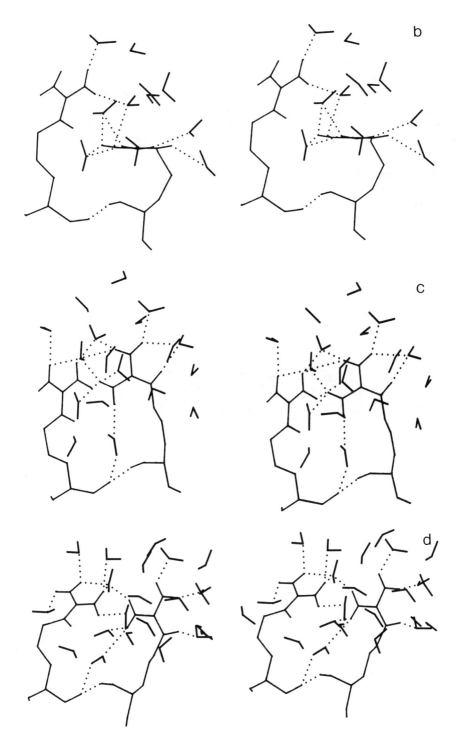

fluctuations are illustrated in the figures with $t > 8.0$ ps, where the hydrogen bonds are indicated by dotted lines.

Also of interest are the correlated motions of different groups that may be important in interpreting chemical events but are not evident in the static average structure. The correlated fluctuations of the protein and the ligand, relative to the dynamics average structure, are shown for the CpA and UV complex of ribonuclease A in Fig. 57a and b, respectively. Although the time scale of the simulation is short compared with the observed rate constant of the enzymatic reaction, it is important to remember that the actual catalytic event in a given molecule occurs on a picosecond time scale; that is, the over-all time scale of a reaction can be orders of magnitude longer than an individual transition due to the presence of activation barriers (see Chapt. IV.E). Many of the large correlations (positive or negative) correspond to atoms in close proximity; an example is the correlation of Thr-45 and the pyrimidine

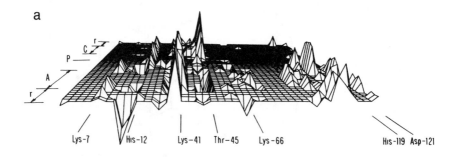

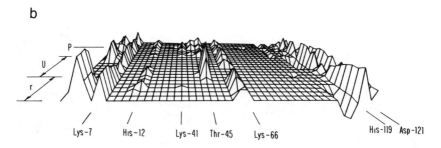

Figure 57. (a) Normalized cross correlations of the atomic fluctuations of the ribonuclease-CpA complex; only correlations with absolute values greater than 0.3 are shown, and an average over all atoms within each protein residue was performed. (b) Same as (a), except for UV complex.

base, which have strong hydrogen bonding interactions. His-119 has large correlations with the phosphate group in CpA and the vanadate group in UV; the latter is of particular interest because there are at least two water molecules, on the average, between the imidazole ring and the phosphate group. An analogous case is Asp-121, which has strong correlations with the substrate in both the CpA and UV complex; in the latter Asp-121 is hydrogen bonded to the CH_2OH group of the sugar, but in the former it is coupled indirectly through His-119 and Phe-120. Lys-7 and Lys-66 are both significantly correlated with the UV ligand, although the N^ϵ groups of the lysines are more than 5 Å removed in both cases. This dynamic coupling appears to be mediated by an intervening water network, whose structural role was described above. By contrast, Lys-41, which interacts directly with the vanadate group, shows only a very weak correlation. In a simulation of the cyclic cytidine monophosphate (CCMP) intermediate analogue, a strong correlation between Lys-41 and the cyclic phosphate group is evident. Such correlated fluctuations between the distant lysine residues and the phosphate group of the substrate may indicate a possible role for the lysines in stabilizing the transition state.

The simulation results suggest several possible functional roles for the solvent and groups of two or more like-charged residues. In the interactions giving rise to specificity and binding, high local densities of charged groups (e.g., positive in the case of ribonuclease A) which are complementary to specific sites on a substrate (e.g., negative in the case of ribonuclese A substrates) will promote binding. Such charge densities are observed to be stable in the presence of strong polar interactions with water molecules. Further, the directing ("steering") of the substrate by the charged groups of the protein into the binding site may increase the rate of binding[155,363,363a] (this chapter, Sect. E below). Also important may be the role of solvent and charged groups in the enzymatic mechanism itself. Possibilities include: (i) a preponderance of charge near a proton-exchange site may favor the reaction;[103,364] (ii) the rates of specific chemical steps may be enhanced by the stabilization of the "transition state," as has been suggested for the active lysine residues in ribonuclease;[103,350] and (iii) the water network and charged groups may help to establish and maintain an active-site conformation that is appropriate for formation of the enzyme-substrate complex.[103]

E. SOLVENT ROLE IN LIGAND-BINDING REACTIONS

The binding of ligands by transport and storage proteins and the binding of substrates and cofactors by enzymes play an essential role in their function. A description of the kinetics of the binding can be divided into several steps. The first step is the approach of the ligand or substrate to the surface of the protein. Once the ligand is near the surface, larger scale protein motions (e.g., hinge bending) start to play a role. As the ligand penetrates into the binding region, details of the protein, ligand, and solvent dynamics can be important. The binding reaction between the ligand and the protein completes the process. In all of these steps, solvent effects can contribute, their role depending in part on the exposure of the region involved.

In the description of oxygen binding to myoglobin (Chapt. VI.B.2), we considered some aspects of the protein motions involved in the penetration step. Solvent was not included in the simulations, although photodissociation experiments have shown that the rebinding process has a significant dependence on the viscosity of the solvent.[318] In the examination of the active site of ribonuclease (see above), we noted the importance of specific water molecules and more generally of solvent extrusion in the binding of substrate. It is clear that for a theoretical analysis of this aspect of the binding process molecular dynamics simulations including the solvent are required. The approach of the ligand to the protein surface is expected to be dominated by the solvent and to be less sensitive to the details of the internal protein motions. In aqueous solution the ligand motion is diffusive and therefore should be amenable to a stochastic dynamic treatment (Chapt. IV.D). We here consider this aspect of the ligand-binding process, as well as some of the larger-scale protein motions that can determine whether the binding site is accessible to the ligand. The escape of the ligand and the removal of product, the rate-limiting step for a number of enzymes (see Chapt. VIII.C.2), can be examined by corresponding methods.

The simplest treatment of diffusive encounters considers two noninteracting hard spheres that have radii r_1 and r_2, diffusion constants D_1 and D_2, and isotropic reactivities such that any collision (i.e., when the distance between the two spheres is equal to the sum of their radii) leads to the reaction. This model requires solution of the Smoluchowski equation[216] with a perfectly absorbing condition at $R = r_1 + r_2$. The resulting rate constant $k(R)$ has the simple form

$$k(R) = 4\pi DR \qquad (90)$$

where $D = D_1 + D_2$. This model has been generalized in a number of ways to include a radial interaction potential between the two spheres,[365] a probability of reaction on encounter that is less than unity,[366] and the possibility that only part of the protein sphere is reactive.[367] All of these generalizations have been introduced in a way that preserves the possibility of obtaining exact or approximate analytic solutions and some have been applied to systems of biological interest.[155,368,369]

To treat more realistic models for the kinetics of biomolecular encounters, a simulation approach has been developed and applied to proteins.[370,371,371a] This approach merges stochastic dynamics methodology (Chapt. IV.D) with the analytic result for the Debye rate constant for a pair of particles, moving diffusively through solvent with a centrosymmetric interaction potential.[365] The analytical expression for the Debye rate constant, $k_D(R)$, to first achieve a separation, R, is given by

$$k_D(R) = \left\{ \int_R^\infty dr \exp[-\beta U(r)]/[4\pi r^2 D_{rel}(r)] \right\}^{-1} \tag{91}$$

where $U(r)$ is the centrosymmetric interaction potential. The reaction space is divided into three spherical regions: $r < R$, $R \le r \le q$, and $r > q$. Only in the innermost region $(r < R)$ are noncentrosymmetric forces important so that the encounter rates for the boundaries R and q can be calculated from Eq. 91; they are $k_D(R)$ and $k_D(q)$, respectively. Stochastic dynamics trajectories are done starting at $r = R$ and it is assumed that reaction occurs if the ligand collides with "active" patches of angular extent θ_0 on the protein at some distance r_0 ($r_0 < R$) from the origin; trajectories that react or for which $r = q$ are terminated. If the fraction of reactive trajectories (i.e., those that collide with the active patches) is β, the rate constant for reaction is given by

$$k = \frac{k_D(R)\beta}{1 - (1 - \beta)\Omega} \tag{92}$$

where $\Omega = k_D(R)/k_D(q)$ corrects for the fact that some of the trajectories that were terminated at the outer boundary $(r = q)$ would have reacted.

Electrostatic interactions, in particular, are likely to influence the encounter dynamics of biomolecular association reactions due to their inherent long-range nature. For example, the charge distribution of an enzyme-substrate system may help draw the two species together and "steer" them into a proper relative orientation for reaction, as suggested for ribonuclease A (this chapter, Sect. D above). An interesting case is the diffusion-controlled transformation of superoxide, O_2^-, to molecular oxygen and hydrogen peroxide by the enzyme superoxide dismutase (SOD). The rate constant for the reaction decreases with increasing ionic strength, despite the fact that both species are negatively charged at neutral pH.[372] It has been suggested that this behavior is due to the presence of positively charged lysine residues in the neighborhood of the active site and their attractive interaction with O_2^-.[363a,373,374] The stochastic dynamics simulation method has been applied to this problem with the SOD dimer and O_2^- modeled as spheres of radius 28.5 and 1.5 Å, respectively.[371] Two reactive patches corresponding to the active-site regions of SOD were defined by surface points within 10° of an axis running through the center of the enzyme sphere. Trajectories were initiated at $R = 300$ Å and terminated upon collision with the active site or with a truncation sphere at $q = 500$ Å. A series of electrostatic models was studied; characteristics of some of the electrostatic models together with calculated rate constants (as values relative to the Smoluchowski rate constant for the uncharged system, Eq. 91) are given in Table X. A dielectric constant of 78 and a solvent viscosity of 1 cP was assumed throughout (hydrodynamic interaction was neglected). The calculated rate for model A, with a negative point charge at the center of the SOD sphere, is somewhat lower than that for the other two models. This supports the idea[363,373,374] that the charge distribution of the enzyme steers superoxide toward the active site. However, the rate calculated with the higher moments included (models B or C) is only two-thirds of that for the same enzyme model in the absence of charges. Models B and C yield essentially the same rate, even though the latter approximates the full charge distribution, and the former reproduces only the first few moments. This corresponds to results obtained for ribonuclease A,[375] where the electrostatic potential seen by the substrate is very similar for a simplified and a complete model for the charge distribution, except very close to the enzyme. Since most of the diffusion time involves large distances, the simplified model provides a good approximation for the encounter rate constant. A recent study[376] of SOD/O_2^- using the charge distribution of model B, but including the dielectric discontinuity at the protein-solvent interface, calculates an increase of the rate, by 5 to 10%, relative to the uncharged reactive patch model; this is an enhancement of 50% over that obtained in the previous study.[371]

TABLE X
Reduced Rates and Electrostatic Models for SOD O_2^- Encounter Kinetics

Model	Electrostatic Characteristics	$k/4\pi D_{rel} R^a$
	No charges	0.066
A	Single point charge of -4 at center of 28.5-Å spherical "enzyme"	0.056 ± 0.004
B	Five point charges which reproduce the monopole, dipole, and quadrupole moments of the SOD dimer	0.079 ± 0.005
C	2196 partial charges assigned to all nonhydrogen atoms of SOD dimer	0.080 ± 0.006

^{a} The distance R is set equal to 30 Å.

Source: Ref. 371.

Use of the Debye-Hückel theory for the electrostatic interactions with model B makes it possible to explore the effects of salt concentration on the rate of encounter.[155] The calculated reaction rate first increases and then decreases to a plateau as the solvent ionic strength increases. The initial behavior at low salt concentration can be attributed to a screening out of the repulsive net charge (monopole) interactions; at higher salt concentrations, the trend is reversed because the shorter-ranged quadrupole moment, which attracts and "steers" the O_2^-, is screened. The experimental data show only the decrease in rate with increasing ionic strength,[372] but it is not clear that a low enough ion concentration was used in the measurements for the monopole screening to be dominant. The improved study[376] as a function of ionic strength indicates that the binding decreases below the rate in the absence of salt above $0.1M$, which is in agreement with experiment.[372] The rate is found to be a maximum near $10^{-3}M$. This behavior is observed regardless of the dielectric model used but has not as yet been verified experimentally.

Internal motions of a protein may influence the details of the ligand binding, particularly the steps involving entrance into and binding to the active site, as well as motion through a protein matrix (Chapt. VI.B.2). The probability that a ligand will be bound upon collision with the protein surface can be less than unity and may fluctuate with time. The character of the relevant motions depends on the particular protein and ligand involved. In systems where the hinge-bending motions of globular domains determine the accessibility of the binding site (see Chapt. VII.B), simplified models for the kinetics can be used to examine the rate constant. It is possible to approximate the protein fluctuations (hinge-bending motions) in terms of a "gate" that regulates access to the binding site. The effects of gating can be introduced into the Smoluchowski treatment or related formulations of diffusion-controlled reactions of particles by means of sink functions or boundary conditions whose intrinsic reactivity fluctuates with time.[154,249,250] In a one-dimensional-like treatment (i.e., a spherical protein and a spherical ligand), the rate constant can be written

$$k = \langle 4\pi R^2 k_s h(t) \rho(R, t) \rangle \qquad (93)$$

where $\rho(R, t)$ is the density of ligands at the ligand-protein contact distance R at time t, k_s is the specific rate constant in the open gate (reactive) state, and $h(t)$ is a characteristic gating function which fluctuates between values of 0 (gate closed) and 1 (gate open). The brackets in Eq. 93 indicate a time average. Thus, the effects of gating on k are found to depend on factors such as the typical lifetimes of the gate-open and gate-closed states and the net rate of motion of the ligand relative to the protein. In the limit of slow gate dynamics, k is just the rate constant for the gate-fixed-open case, multiplied by the fraction of the time that the gate is open. More complex cases involving different forms for $h(t)$ that lead to substantial deviations from this "intuitive" result have also been examined.[154,249] A general treatment has been developed that makes it possible to include a gating mode in the presence of anisotropic protein and ligand reactivities.[250] Of interest is the conclusion that the coupling of the orientationally restricted reactivities and the gating mode leads to a significantly larger reduction of the rate than superposition of the separate effects.

CHAPTER X

THERMODYNAMIC ASPECTS

Understanding proteins and the details of their functions requires an evaluation of their thermodynamic properties, as well as of their dynamics. In this chapter we focus on the thermodynamic aspects of structure and reactivity in biopolymers. Emphasis is given to studies that have applied the dynamical methodologies described in Chapt. V.

A. CONFORMATIONAL EQUILIBRIA OF PEPTIDES

An important question regarding peptides and proteins is concerned with the equilibria among several conformational states. It has been suggested that enzyme function may be linked to the occurrence of particular conformations in solution.[24,377] A mechanism recently proposed for the hydrolysis of oligosaccharides by the enzyme lysozyme, for example, is based on the observation of specific substrate and enzyme sidechain conformations in a molecular dynamics simulation of a lysozyme-substrate complex.[378] Also, local conformational equilibria and the barriers between conformations are important in determining the rates and mechanisms of folding and rebinding processes.

The distribution of configurations that exist at a given temperature is determined by the relative free energies. Since most processes of biological interest occur in an aqueous environment, a knowledge of the conformational free energies, or more precisely, free-energy differences, in solution is required. Simulation methods including the solvent explicitly can be used for the determination of free-energy differences although they are very time consuming.[342,379,380] In many cases qualitative and even semiquantitative features of the solvent effect on conformational equilibria can be obtained more simply from the potentials of mean force determined by integral equation methodologies (Chapt. V.C). In addition, simulations carried out on the potential-of-mean-force or free-energy surface provide insights into the importance of fluctuations, including their contribution to the configurational entropy. An application to a study of the conformational equilibria in a small biopolymer, the dipeptide N-methylalanyl acetamide, will be used to illustrate this approach.

Figure 58 shows a comparison of the potential energy map (Ramachan-

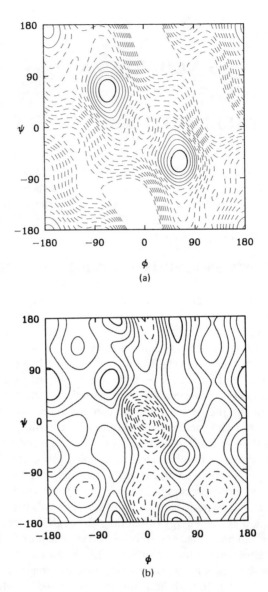

Figure 58. Energy contour map (Ramachandran plot) for ϕ versus ψ of the alanine dipeptide. Solid lines mark the five lowest-energy contours at 1 kcal/mol and the bottom contour is marked with a heavy line; dashed lines mark the higher contours at 1-kcal/mol intervals: (a) vacuum potential surface; (b) solvent-modified potential surface.

dran plot) for the ϕ and ψ angles in the "alanine" dipeptide in vacuo (Fig. 58a) and in an aqueous environment (Fig. 58b) obtained by use of integral equations.[115] This figure demonstrates the important influence that water can have on the relative stabilities of different conformers. In vacuum the C_{7eq} ($\phi = -67°$, $\psi = 65°$) and C_{7ax} ($\phi = 63°$, $\psi = -62°$) conformations are the only ones that are significantly populated due to the presence of an internal hydrogen bond. In aqueous solution this is no longer true. Instead, a much wider range of conformations is found to be accessible. This is illustrated most dramatically by Boltzmann–weighted probability maps (Fig. 59a and b) calculated from the Ramachandran plots.

From a comparison of the two surfaces it is evident that the barriers between the C_{7eq} conformation and the α_R ($\phi = -69°$, $\psi = -48°$) or the P_{II}

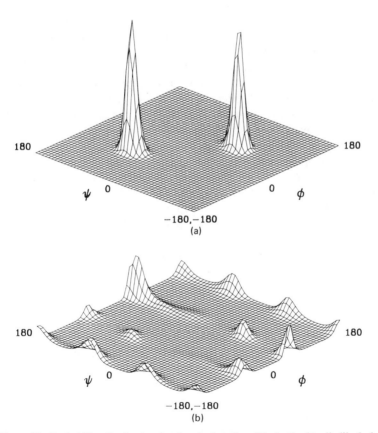

Figure 59. Probability distribution for the alanine dipeptide in the (ϕ, ψ) dihedral-angle space at 300 K: (a) vacuum potential surface; (b) solvent-modified potential surface.

($\phi = -55°$, $\psi = -176°$) conformations are substantially lowered in water. The α_R and P_{II} conformations, which are not present as minima on the vacuum surface, are minima in water. The C_5 conformation (near $\phi = -176°$, $\psi = 172°$) and additional helical geometries are slightly stabilized relative to the C_7 conformers in solution. A Monte Carlo simulation of the dipeptide[342] has determined the solvent contribution to the free-energy differences among the three important conformations α_R, C_{7eq}, and P_{II} by use of umbrella sampling. Although a quantitative comparison is not possible because different potential functions were used, the α_R and P_{II} conformers are stabilized relative to C_{7eq} by 3.6 and 3.2 kcal/mol, while the integral equations yield 8.8 and 10.0 kcal/mol,[115] respectively, for a purely electrostatic hydrogen-bond model and 5.2 and 4.7 kcal/mol, respectively, for a model including explicit (nonelectrostatic) hydrogen-bond functions.[380a]

The change in the free-energy potential surface (potential of mean force surface) induced by solvent is expected to have an effect not only on the relative free energies of the different minima but also on the dynamics and fluctuations of the molecules in the various minima.[115,380] By use of a molecular dynamics simulation and the quasi-harmonic method (see Chapt. IV.F), the differences in the internal vibrational entropy of different conformers of the alanine peptide have been compared in vacuum and in solution.[380] It was found for the C_{7eq} conformation that the vibrational entropy change is -1.41 cal/mol-K, corresponding to a net entropic destabilization of 0.42 kcal/mol at 300 K in going from vacuum to solution. Comparison of the vibrational free energies of different conformers in solution showed that P_{II} is stabilized relative to C_{7eq} by 0.48 kcal/mol at 300 K, while α_R and C_{7eq} have nearly the same internal vibrational free energy.

It is clear from this simple example that water can have a profound influence on conformational equilibria of small peptides. Corresponding effects are expected for other solvated biomolecules and individual exposed amino acid sidechains. The integral equation method is likely to be very useful in studying the effects of solvation on the conformational states of small drug and substrate molecules and to provide insights into the role of flexibility on the binding affinity of ligands.[26a]

A logical extension of conformational free-energy surfaces for small peptides is the evaluation of the secondary structural propensities of larger systems. A difficulty encountered in the study of polypeptides is the fact that the number of extrema in the free-energy surface rapidly becomes intractable with increasing size. However, there exist empirical procedures for assessing the likelihood that a given sequence of amino acids will adopt a certain type of secondary structure in a protein. Although these are phenomenological constructs, rather than theories for the relative free energies, they have been used to organize and correlate the available data. In addition, some insight into the problems of predicting secondary structure from theoretical calculations may be obtained from the constructs themselves. The earliest methods did not make use of protein structures but relied on the optical rotary dispersion (ORD) spectrum of different secondary structural types.[381,382] Given the ORD spectrum and the primary sequence of amino acids, correlations between the helicity and the types of residues emerged. With the advent of protein crystallography, the accuracy of the predictive methods could be tested. The methods were found to overpredict the amount of either α-helix or β-sheet content in a given sequence (see Chapt. II.A). Based on known crystal structures, a variety of rather intricate statistically based procedures have emerged for the prediction of secondary structures.[383] These empirical rules yield results that are better than the random for the distribution of β-sheets, β-bends, α-helices, and random coil regions when applied to sequences of known structure. However, their accuracy is not sufficient to predict correctly the entire secondary structure of most proteins. The desire for a more basic understanding of secondary structural determinants, intermediate between the explicitly atomic theories and the empirical rules, has led to statistical mechanical theories[384] based on residue interactions.

Extensions of the empirical, simulation, and integral equation methods to globular proteins and nucleic acids may provide a link between conformational equilibria, stability, and biological function. It is an essential element of the dogma of protein folding that the effect of solvent is paramount in such equilibria.[316] This has not yet been verified in a quantitative manner either experimentally or theoretically. It is known, however, that dry proteins (i.e., film aggregates in the absence of water) have essentially the same structure as the native form in solution[385] and that with even a small amount of water (e.g., for lysozyme approximately 130 water molecules per protein), proteins function normally.[386] Of course, the film experiments may involve metastable species (perhaps analogous to those simulated in vacuum molecular dynamics studies), but the data do give one pause about blindly accepting the dogma concerning the role of solvent.

Interactions between specific solvent molecules and protein atoms can be important in protein function and stability. Many enzyme reactions, for example, involve the participation of water molecules. To study such effects, a methodology that retains solvent molecules explicitly is required. Both conventional molecular dynamics techniques and the stochastic boundary molecular dynamics approaches can be used. However, when the region of interest is localized, or energy exchange is expected to be important, the stochastic boundary methods are more convenient. In simulation studies of the active-site regions of native ribonuclease A[103] and lysozyme[108] it was observed that water molecules form hydrogen-bonded networks which stabilize several charged residues in the active-site region; in some cases these networks were extensive. Of particular interest is the fact that the water network is capable of stabilizing sidechains in configurations with like-charged groups [e.g., $(NH_3)^+$ of Lys, $(NH_2)^+$ of Arg, and (NH^+) of His] in close contact (typical $N \cdots N$ distances of about 3.5 Å were observed). This phenomenon is likely to be of general significance, although it is mentioned only briefly in the experimental literature. Active-site stabilization of like charged groups by water molecules has been described in Chapt. IX.D.

B. CONFIGURATIONAL ENTROPY OF PROTEINS

It is common practice[47,387] to estimate the configurational entropy change in protein denaturation, ΔS_{conf}, by setting the configurational entropy of the native protein equal to zero and calculating that of the random-coil state from the number of possible conformations, generally taken to be equal in energy for simplicity. This leads to a value for ΔS_{conf} in the range 4 to 6 cal/(mole of residue-K) and to a room-temperature contribution to the free energy of denaturation of 1.2 to 1.8 kcal per mole of residue. Experimental values of the denaturation entropy at ambient temperatures are in this range; there is a significant temperature dependence due to the large value of the heat capacity of denaturation.[387]

Using the methods for treating the dynamics of proteins described in Chapt. IV, it is possible to estimate the residual configurational entropy due to the fluctuations of the folded polypeptide chain. To calculate this property, both molecular and harmonic dynamics approaches can be used.[136,140,141] Although the harmonic approximation yields an overestimate of the classical entropy,[140] it is useful because the essential quantum corrections can be introduced most easily (Chapt. V.A). For the well-studied protein BPTI in its native state the residual entropy, S_p^v, evaluated from the normal-mode frequencies and the quantum-mechanical partition function (Eqs. 44 to 47) is found to be 2006 cal/(mol-K) or 34 cal/(mole of residue-K). This corresponds to a free-energy contribution arising from the internal motions of the folded protein equal to -617 kcal/mol or -10.6 kcal per mole of residue at 300 K. In a normal-mode calculation for BPTI that included only dihedral degrees of freedom,[136a] a much smaller value for S_p^v, 1066 cal/(mol-K), was obtained. This demonstrates that important contributions arise from degrees of freedom other than the torsions (e.g., coupling between bond angles and dihedral angles)[380] and suggests that these must be included in evaluating thermodynamic properties.

Of interest also is the heat capacity, which is sensitive to the density of states in the frequency range of interest. Although no experimental results for BPTI are available, heat capacity measurements have been made for a number of proteins in solution.[388] They yield a room temperature heat capacity

(C_p) equal to 0.32 ± 0.2 cal/(K-g). This can be compared with the calculated vibrational C_v for BPTI, which equals 0.25 cal/(K-g). The calculated value of dC_v/dT for BPTI at 300 K is 0.0011 cal/(K^2-g); the measured range for various proteins is 0.001 to 0.002 cal/(K^2-g). These comparisons suggest that the internal entropy of BPTI obtained from the harmonic model is of the right order of magnitude.

The fact that the calculated value of S_p^v, normally set equal to zero, is nearly an order of magnitude larger than the estimates of ΔS_{conf} suggests that the difference between the folded and denatured state should be considered in more detail. In a denatured protein, which is assumed to approximate a random-coil polymer, there are two contributions to the configurational entropy.[140] The first is that due to the local fluctuations in the neighborhood of a well-defined structure and the second corresponds to the existence of more than one such structure. In the standard treatments the former is neglected and only the latter is included. To illustrate what happens when both contributions are considered, a model is used in which a molecule has N conformations, each of which can be treated as a disjoint harmonic well.[389] The total configurational entropy, S_{conf}, can be written

$$S_{conf} = \sum_{I=1}^{N} \omega_I S_I^v - k_B \sum_{I=1}^{N} \omega_I \ln \omega_I \qquad (94)$$

where ω_I is the Boltzmann weighing factor for the zero of energy of well I and S_I^v is the vibrational entropy of well I; S_I^v can be calculated with the harmonic model by use of Eqs. 44 to 47.

In Eq. 94, the second term, the so-called entropy of mixing, is the one usually equated to the total configurational entropy of the denatured state. Under the assumption of equal weights ($\omega_I = \omega$), it reduces to the standard expression, $k_B N \ln N$, for the entropy. The first term in Eq. 94 represents the Boltzmann-weighted sum of the configurational entropies of the individual conformers. The denatured state has contributions from both terms in Eq. 94, with the sum extending over the N allowed conformations. The entropy of the folded protein can also be described by Eq. 94. In the simplest approximation for a native protein, only the first type of term is present; i.e., there is a single conformation with residual entropy S_p^v (but see below), associated with the vibrational degrees of freedom, and it is this term that was evaluated above for BPTI.

The configurational contribution to the denaturation entropy can now be written

$$\Delta S_{conf} = S_{rc} - S_p^v = \left(\sum_{I=1}^{N} \omega_I S_I^v - k_B \sum_{I=1}^{N} \omega_I \ln \omega_I \right) - S_p^v \qquad (95)$$

where S_{rc} is the entropy of the denatured (random-coil) state and the sum over I is over the N conformers of the denatured state. Since the calculation for BPTI shows that $S_p^v \gg (-k_B \Sigma \, \omega_I \ln \omega_I)$, the usual approximation for ΔS_{conf} requires that

$$\sum_{I=1}^{N} \omega_I S_I^v \simeq S_p^v \qquad (96)$$

To examine this possibility the vibrational entropy of individual conformers of a number of blocked amino acids with the same potential function as used for the protein has been calculated.[389] The values of S_I^v were found to range from 19 cal/(mol-K) for Ala to 48 cal/(mol-K) for Trp with an average of 34 cal/(mol-K). The exact values depend on the nature of the blocking group, but the order of magnitude should be valid. For decaglycine in a single extended configuration,[140] the vibrational entropy was found to have a similar value on a per residue basis [28 cal/(mole of residue-K)].

These results suggest that the fluctuations, and therefore the configurational entropy of a folded protein, are rather similar to those of a random coil in a single potential minimum. This leads to the following argument. The vibrational entropy of a protein is approximately an extensive property; i.e., a protein is large enough so that $S_p^v = n\langle S_{res}^v \rangle$, where $\langle S_{res}^v \rangle$ is the average over the different amino acids residues of the configurational entropy for a single conformation [i.e., $\langle S_{res}^v \rangle \simeq 34$ cal/(mole of residue-K)]. This conclusion finds support in a comparison of calculations for BPTI and lysozyme.[389] If $\langle S_{res}^v \rangle$ is the same in the native protein and for all configurations of the denatured state, Eq. 96 is satisfied and

$$\Delta S_{conf} \simeq -k_B \sum_{I=1}^{N} \omega_I \ln \omega_I \qquad (97)$$

as in the standard model.

An additional factor to consider is the magnitude of the anharmonic contribution to the entropy of proteins. Analyses of molecular dynamics simulations have demonstrated that the major anharmonic contributions can be ascribed to multiple conformations for individual atoms.[196] In BPTI and lysozyme, an estimate of the change in entropy due to these effects (e.g., 89 atoms in lysozyme have multiple wells) yields a correction of less than 2% for the classical entropy. Thus multiple conformations appear not to be important for the residual entropy at ordinary temperatures.[389] However, near absolute zero (1 to 2 K) there are data that suggest that several minima ("tunneling states") contributes significantly to the entropy.[390]

The conclusion that the $\langle S_{res}^v \rangle$ is essentially the same for a protein in its native conformation and for a single conformer of the denatured polypeptide chain rationalizes the agreement between simple model calculations with

$S_p^v = 0$ and the measured denaturation entropies. However, the large magnitude of S_p^v raises some quantitative questions. These are likely to be most important for small perturbations of a protein where a quantitative understanding of the entropy changes is required. One such case involves ligand binding,[29a,208,391] for which it has been suggested that the distribution of vibrational frequencies is altered in going from the unliganded to the liganded protein; e.g., for BPTI some change in the configurational entropy might be expected on binding to trypsin. As a model for this effect, a comparison is made of the results obtained in BPTI from the frequencies calculated from the normal-mode analysis and from a set of adjusted frequencies,[136] the latter representing the perturbed system with higher frequencies. At 100 K, the vibrational free energy changes from −41.5 kcal/mol to −37.7 kcal/mol in the presence of the perturbation; at 300 K, the values are −336.4 and −325.1 kcal/mol, respectively. At all temperatures the vibrational energy increases while the entropy decreases, leading to a significant destabilizing effect of the system.[29a] The change in energy contrasts with that found in previous qualitative discussions,[208] due to the fact that the zero-point contribution was neglected in the latter.

A case of recent interest concerns the effects of single-site mutations on the stability of proteins. The most specific data are available for T_4 lysozyme mutants.[392,392a] Several single amino acid substitutions have been studied that leave the crystal structure essentially unaltered within the accuracy of the X-ray data. The observed changes in the denaturation entropy at 320 K vary from −59 to +5 cal/(mol·K), relative to the wild-type value of 236 cal/(mol·K); the changes in the denaturation enthalpy range from −21 to 0.6 kcal/mol at the same temperature with considerable enthalpy-entropy compensation. That such large changes in the thermodynamic properties are introduced by altering one residue does not appear unreasonable in view of the calculated per-residue entropy of 34 cal/(mol·K). However, specific calculations (e.g., with thermodynamic perturbation theory (Chapt. V.C)) are required to obtain a detailed explanation of each mutant. In this regard, it is important to note that by the nature of the measurements (i.e., differences in thermodynamic quantities are being considered), it is possible that the observed values for the mutants arise from changes in the properties of the native state, the denatured state, or both. An analysis of mutant data for staphylococcus nuclease suggests that the properties of the denatured state play an important role.[393]

C. LIGAND BINDING, MUTAGENESIS, AND DRUG DESIGN

Thermodynamic perturbation theory (Chapt. V.C) provides a convenient approach to a variety of problems concerned with the free energy of binding. More generally, one can use thermodynamic perturbation theory to determine relative free energies between different chemical or thermodynamic states of systems of interest. A well-defined example is provided by the problem of the difference in the binding free energy of a substrate with a hydrogen at a given site and one with a methyl group at the same site (e.g., in the substitution of glycine by alanine). Here the reference system is the substrate with the hydrogen and the perturbation coordinate corresponds to the metamorphosis of the hydrogen into a methyl group. In the simplest model, which treats the methyl group as an extended atom, it is possible to use Eq. 64 and to integrate along the path that gradually changes the radius and well depth of the hydrogen atom into that of a methyl group. This yields the binding free-energy difference between the two substrates. By going through the metamorphosis procedure for the substrate in aqueous solution and at the binding site, one can determine which substrate has the greater binding free energy in the solution environment (see Fig. 10c). It is this quantity (referred to as $\Delta\Delta A$) that is of primary interest because it is directly related to the measured difference in binding constants. In this type of calculation, the implicit assumption is made that the drug binds to the receptor in the same way with a hydrogen and a methyl group; if it does not, a more complicated thermodynamic integration procedure is required. For large and complex structures, simulation methods are required to obtain the information necessary for the perturbation calculation. In simpler cases, integral equation theories can be used with a great reduction in computational effort.

When using simulation techniques and a linear scaling of the perturbation potential is appropriate (see Eq. 52), the quantity in the integrand of Eq. 64, $\langle V_\lambda(\mathbf{r}^N)\rangle_\lambda$, is the mean potential energy of the simulated system as a function of λ. Since this can be evaluated in a straightforward manner, such free-energy simulations provide a powerful method for calculating many features of interest of protein-substrate-solvent systems. In the most detailed treatment all the atoms composing the substrate, the protein, and the solvent molecules (e.g., a system with periodic boundary conditions) are included explicitly and a trajectory in phase space is determined for this many-particle system to obtain the average potential energy for a given value of λ. Either molecular dynamics or Monte Carlo methods can be used in this approach, although both calculations are time consuming. The essential problem is that for efficient sampling it is necessary to restrict the integration interval to small changes in the system. If the change of interest is large (and even the transformation of a hydrogen to methyl group is a "large change" in this sense), it is necessary to carry out a series of successive simulations; e.g., if Eq. 64 is being used and the comparison involves $\lambda = 0$ (a hydrogen atom) and $\lambda = 1$ (a methyl group), a series of intermediate simulations ($\lambda = 0, 0.1, 0.2, 0.3$ etc.) may be required to determine the overall free-energy change.

Early simulations by thermodynamic perturbation theory include an evaluation of the free energy of liquid water[394] and the free energy of spherical cavity formation in liquid water;[174] the latter created the solvated atom by starting with a zero radius and increasing it in a stepwise or nearly continuous fashion to the appropriate value. A thermodynamic cycle in conjunction with perturbation theory has been used for a primitive model of a ligand-receptor-solvent system.[395] The solvent consisted of an equal number of slightly positive and negative charged Lennard-Jones spheres. The "receptor" was a fixed sphere with a larger charge and the "substrate" was an atom with a charge opposite to that of the receptor. The necessary averages were performed by a molecular dynamics simulation. By using the charge as the perturbation parameter, the difference in free energy of binding between a neutral ligand ($\lambda = 0$) and a charged ligand ($\lambda = 1$) was determined. As an alternative, the physically meaningful thermodynamic cycle was followed (see Fig. 10) by means of umbrella sampling for the ensemble averaging and comparable results were obtained. A more realistic calculation[396] employed a corresponding approach to compute the relative free energy of binding for Cl^- and Br^- to the macrocyclic ionophore, SC-24, in aqueous solution. The thermodynamic perturbation calculation gave a relative free energy of binding for Cl^- versus Br^- of 4.1 kcal/mol, which compares favorably with the experimental difference of 4.3 kcal/mol. However, a quantitative evaluation is difficult because although the relative free energy of binding has been measured, there is no accurate experimental value for the difference in free energy of solution between Cl^- and Br^-.

An analogous approach has been applied to the binding of benzamidine,[379,397] an inhibitor, to the enzyme trypsin. A full periodic boundary simulation was done with 16,384 atoms that included the aqueous solvent explicitly. In one study the relative change in free energy between benzamidine and *para*-fluorobenzamidine was calculated. A value of 0.9 ± 0.5 kcal/mol at 300 K was obtained. This may be compared to the experimental estimate of 0.5 kcal/mol (approximately corrected for the appropriate thermodynamic state). Although the experimental comparison is not unequivocal, the set of calculations has led to interesting mechanistic insight into the binding. By considering the various contributions in the thermodynamic cycle depicted in Fig. 10c, it was found that the major effect in the difference in binding for benzamidine versus the fluorinated compound arises from the desolvation contribution, not the specific enzyme-inhibitor interactions. This is analogous to the results obtained in binding energy calculations of netropsin and a series of related compounds to the minor groove of B-DNA.[398] By performing another series of perturbation calculations,[379] the change in free energy upon binding of benzamidine to trypsin relative to a mutant trypsin was estimated. Specifically, Gly-216 was perturbed into Ala-216. This amino acid residue is in the region involved in inhibitor binding, so that there is a free-energy penalty due to the interaction between the protruding methyl group and the benzamidine in the active site of the mutant enzyme. The calculated free-energy difference of 1.3 kcal/mol is similar to a value (~ 2.0 kcal/mol) estimated from kinetic experiments for a homologous system. A number of similar studies on other systems have now been published.[398a-399]

Related problems that can be probed with such perturbation techniques involve the determination of the free energy along the reaction coordinate for a chemical reaction in solution. An example is provided by the symmetric chlorine exchange (S_N2) reaction

$$Cl-CH_3 + Cl^- \rightarrow Cl^- + CH_3-Cl \qquad (98)$$

Making use of the interaction potential between the solute molecules involved in the reaction as obtained from quantum mechanical calculations, the free-energy change along the reaction coordinate in aqueous solution has been calculated and compared with the gas-phase result.[399] Equation 52 and a Monte Carlo simulation with umbrella sampling were employed. The result obtained is in good agreement with the experimental data[400] for this system. Similar results have been determined with a combined quantum and classical mechanical simulation method that employed the λ integration scheme (Eq. 59) rather than umbrella sampling.[132b] The theoretical treatments have given insight into the role of solvent stabilization for this reaction. It was found that the ion-dipole minimum in the gas-phase potential surface has almost disappeared in aqueous solution.[399] Thus the activation energy of the transition-state complex for this reaction is very different in the gas phase and in solution. An integral equation calculation for the same reaction obtained similar results.[179] Free-energy simulation methods have also been applied to the reaction catalyzed by trypsin to investigate the effect of the same mutation considered in the benzamidine binding discussed above. Approximate agreement was obtained for the change in the rate constant for hydrolysis induced by the mutation.[401] A more general analysis of the rate enhancement by trypsin has also been made by simulation methods.[402]

The interplay between theory and experiment in this area is exemplified by studies of the affinity of myoglobin for xenon. High-resolution crystallographic investigations have indicated that there are several hydrophobic cavities in myoglobin.[403-405] When myoglobin is subjected to an overpressure of xenon, the most highly occupied site corresponds to the cavity located on the distal side of the heme.[404,405] This is in accord with NMR measurements.[406] Most of the cavities are in the interior of myoglobin with no clear path to the surface. Thus the situation is similar to that already discussed in Chapt. VI.B.2 for oxygen and carbon monoxide binding to the heme group. Simulations have been done to find the barriers for exit and entrance of the xenon atoms. To estimate the free energy of binding for xenon at the distal site, a thermodynamic integration has been performed with the stochastic boundary molecular dynamics method.[407] After the usual relaxation and thermal equilibration of the protein, a continuous variation in λ was used to sample $\langle V_\lambda (\mathbf{r}^N, \lambda) \rangle_\lambda$ over the interval $0 \leq \lambda \leq 1$; as in the hydrophobic solute simulation, the parameter λ corresponded to the van der Waals radius of the atom. The integration in Eq. 64 is then simply performed (see Chapt. V.C). As a check of convergence of the computational procedure, the sampling was performed in reverse; that is, the initial conditions consisted of the xenon already in place and equilibrated at the distal site and λ was slowly decreased from 1 back to zero. The forward integration yielded 5 kcal/mol and the reverse integration differed by only 0.5 kcal/mol. Although this is only a lower bound on the possible error, the near reversibility of the calculation does imply that complications arising from essentially irreversible structural relaxations (on the time scale of the computer experiment) are not important. The experimental estimate of the binding energy for xenon at the distal site is 4 kcal/mol.[406]

All of the thermodynamic simulations that have been reviewed in this section resulted in calculated free-energy changes that are in qualitative or semi-quantitative agreement with experiment. Since one would like to use the thermodynamic integration technique for predictive purposes, it is essential to introduce some cautionary remarks. Because most perturbations considered so far are relatively small, it is not surprising that the calculations yield small free-energy differences and that these are of approximately the right order of magnitude. However, quantitative results, which are of primary interest, are very sensitive to the choice of potential (possibly including the truncation of the long-range interactions) and to the convergence of the method used to evaluate the ensemble averages. It appears that the precision of the perturbation calculations is very high[407a] but concerning their accuracy there is still considerable uncertainty. Thus care must be exercised in evaluating the successes of the method. In a test of thermodynamic integration for the free energy of hydration of methanol in water,[408] performed by transforming water to methanol, a range of values between -0.5 and $+3.6$ kcal/mol was obtained for a series of different but reasonable potential functions; the experimental value is $+1.2$ kcal/mol.[409] Also, in a simulation of the difference in binding free energy of two different inhibitors to dihydrofolate reductase, the right magnitude (1.4 kcal/mol) but wrong sign was calculated.[409] This suggests that at the present stage of refinement, comparisons of calculated free-energy results with experiment can serve as useful tests of simulation methods and potential functions. Hopefully, the reliability will be improved to permit meaningful predictions as well.

Although simulations are, in principle, the most appropriate technique for the evaluation of relative thermodynamic properties of macromolecular systems, energy minimization will continue to be an important tool for comparing the binding of related compounds with less computational effort than is required for a free-energy calculation. Equations 42 and 43 can be used to make estimates of the binding energy. An example is provided by a study of the Michaelis complex and tetrahedral intermediate of N-acetyltryptophanamide with α-chymotrypsin, which has been examined for stereoselectivity.[410] This is a particularly good case for a molecular mechanics study because the L and D forms of the ligand have the same solvation energy and free energy. Starting with the X-ray structures of the native enzyme and that with a tosyl inhibitor, a series of positions for the bound L and D ligands were generated and their energies calculated with an empirical force field (Eq. 6). No selectivity for a given stereoisomer of the substrate was calculated without the incorporation of flexibility in the enzyme-substrate complex through the use of a conjugate gradient minimization (i.e., Eq. 43 rather than Eq. 42 had to be used for the calculation). Hydrogen-bonding interactions with His-57 and Ser-214 were reported to be responsible for the differentiation between the D and L isomers in the transition-state calculation. Although the order of the energies obtained for the tetrahedral intermediate of the L and D forms is in agreement with experiment, the absolute values are much too large. As is pointed out,[410] neglect of solvent and the internal mobility of the system are drastic approximations in the model, which has nevertheless provided useful insights. It would clearly be of interest to apply free energy simulations to this system, although possible differences in the binding modes for the D and L forms of the substrate increase the difficulty of the calculation.

So far consideration has been limited to systems for which the structures of all the molecules involved in the calculation are known. It is of more than theoretical interest to be able to develop structure-activity relations and to identify pharmacophores (the essential common elements of a series of drugs that have the same or similar effects) in ligands for which no structural data concerning the enzyme or receptor exist. Areas of biochemistry where such work is important include the study of hormones, neuropeptides, and other psychoactive compounds. A common theme of the techniques used in studying the binding at sites of unknown structure is the correlation of the results from a series of related molecules known to be active and specific to a particular receptor.[410a,410b] Such information allows one to place conditions on the design of new drugs. The techniques described in Chapt. IV, particularly energy minimization, are generally employed in this type of molecular modeling.

Research in the area of drug design is inspired by both fundamental and practical concerns.[410c] Examples of practical applications are the modification of drugs to reduce undesirable side effects or to increase activity or specificity. Some of the more basic studies are concerned with understanding the conformational properties of peptides or other ligands in solution and the relation of their solution free-energy states and kinetics to those involved in the drug-receptor binding.

An approach that combines energy minimization, the computation of electrostatic potential surfaces, and interactive computer graphics has been applied[411] to a family of clozapine-like neuroleptics. The potent antipsychotic activity of clozapine, once overlooked as a neuroleptic, is now well documented; however, its hematological toxicity is a very undesirable side effect. Thus a goal of the work was to identify the pharmacophores common to clozapine and related compounds that showed the neuroleptic properties without the undesirable side effects. Minimization studies were done with an empirical potential starting with initial coordinates based on X-ray crystallographic data for the ligand. The electrostatic potential on the solvent accessible surface was computed from charges assigned by quantum-mechanical calculation. A graphics system was then used to examine the electrostatic potential surface.[411] The results of the pharmacophore search were found to be relatively insensitive to the exact method used to determine the partial charges. Also, clozapine and its des-chloro analog displayed similar pharmacological profiles and similar electrostatic surfaces. Other compounds which retained the neuroleptic activity but lacked the side effects showed markedly differing electrostatic surfaces. This suggests the somewhat surprising result that the overall electrostatic potential, rather than specific heteroatom substitutions, is important for the pharmacological properties of this family of compounds. As yet no new active compounds have been synthesized as a consequence of this study.

Both energy minimization and molecular dynamics have been used to study a group of Met-enkephalin, [H-Tyr-Gly-Gly-Phe-Met-OH] analogues.[412] Cyclized analogs such as [d-Pen^2, d-Pen^5] enkephalin, where Pen designates the modified amino acid penicillamine, have been synthesized and characterized pharmacologically. The bis-penicillamine compound was found to be both active and specific with respect to certain of the binding site for Met-enkephalin. Even with the cyclized ring, energy minimization yielded a large number of minima on the potential energy surface. By use of data from NMR studies, many of the minima could be eliminated. A molecular dynamics simulation was then made at elevated temperatures (1000 K) to facilitate the search for other minima. Although no unexpected minima were uncovered, a hindered crankshaft-like motion involving the disulfide dihedral angle was observed to occur in certain structural transitions. The motion of the disulfide dihedral is hindered by the presence of geminate dimethyl groups which are part of the Pen residues.[412] This particular motion may be of biological significance, as it also appears in several related compounds which have only slightly lower activity and specificity. The relative orientation of the two aromatic rings is not affected by the S—S dihedral motion. A hypothesis resulting from the simulation is that instead of the previously suggested one-aromatic-ring pharmacophore, both the Phe and the Tyr rings may be part of the binding and recognition moieties. In more highly constrained inactive compounds, such as the bis-penicillamine tetra-hydroisoquinoline analogues, the crankshaft motion interferes with relative ring orientation. This is true even though there are stable minima corresponding to structures which do fit the pharmacophore. Although these hypotheses, including the possible role of internal motions, are suggestive, additional experimental work is required to determine their validity for the opiate-receptor-binding problem.

CHAPTER XI

EXPERIMENTAL COMPARISONS AND ANALYSIS

This chapter focuses on experimental methods that have a symbiotic relationship with theoretical studies of proteins. In most cases, the need for simulations arises because the complexity of the internal motions that take place in proteins is such that the simple models used to interpret corresponding experiments on small molecules are not necessarily appropriate. From an analysis of the simulations, more realistic models can be constructed and parametrized in such a way that they can be fitted to the experimental data. Simulations can also serve to test approximations in the methods used to interpret the experiments and to provide additional information that makes it feasible to use the measurements in a new and interesting way. Examples of such applications of simulations to improve experimental analyses are given in this chapter.

A. X-RAY DIFFRACTION

Since atomic fluctuations are the basic elements of the dynamics of proteins (see Chapt. VI.A), it is important to have experimental tests of the accuracy of the simulation results. For the magnitudes of the motions, the most detailed data are provided, in principle, by an analysis of Debye-Waller or temperature factors obtained in crystallographic refinements of X-ray structure.

It is well known from small molecule crystallography that the effects of thermal motion must be included in the interpretation of the X-ray data to obtain accurate structural results. Detailed models have been introduced to take account of anisotropic and anharmonic motions of the atoms and these models have been applied to high-resolution measurements for small molecules.[413] In protein crystallography, the limited data available relative to the large number of parameters that have to be determined have made it necessary in most cases to assume that the atomic motions are isotropic and harmonic. With this assumption the structure factor $F(Q)$, which is related to the measured intensity by $I(Q) = |F(Q)|^2$, is given by

$$F(\mathbf{Q}) = \sum_{j=1}^{N} f_j(\mathbf{Q}) e^{i\mathbf{Q}\cdot\langle\mathbf{r}_j\rangle} e^{W_j(\mathbf{Q})} \qquad (99)$$

where $\mathbf{Q}$ is the scattering vector and $\langle\mathbf{r}_j\rangle$ is the average position of atom j with atomic scattering factor $f_j(\mathbf{Q})$ and Debye-Waller factor $W_j(\mathbf{Q})$, defined by

$$W_j(\mathbf{Q}) = -\tfrac{8}{3}\pi^2 \langle\Delta r_j^2\rangle s^2 = -B_j s^2 \qquad (100)$$

where $s = |\mathbf{Q}|/4\pi$; the sum in Eq. 99 is over the N atoms in the asymmetric unit of the crystal. The quantity B_j is usually referred to as the temperature factor, which is directly related to the mean-square atomic fluctuations in the isotropic harmonic model. More generally, if the motion is harmonic but anisotropic, a set of six parameters $(B_j^{xx} = \langle\Delta x_j^2\rangle, B_j^{xy} = \langle\Delta x_j \Delta y_j\rangle, \ldots, B_j^{zz} = \langle\Delta z_j^2\rangle)$ is required to characterize the atomic motion fully. Although in the earlier X-ray studies of proteins the significance of the temperature factors was ignored (presumably because the data were not at a sufficient level of resolution and accuracy), more recently attempts have been made to relate to the observed temperature factors to the atomic motions. In principle, the temperature factors provide a very detailed measure of the motions because information is available for the mean-square fluctuation of each heavy atom. In practice, there is a basic difficulty in relating the B factors obtained from protein refinements to the atomic motions. The first is that in addition to thermal fluctuations, any static (lattice) disorder in the crystal contributes to the B factors; i.e., since a crystal is made up of many unit cells, different molecular geometries in the various cells have the same effect on the average electron density, and therefore the B factor, as atomic motions. In only one case, the iron atom of myoglobin, has there been an experimental attempt to determine the disorder contribution.[202] Since the Mössbauer effect is not altered by static disorder (i.e., each nucleus absorbs independently) but does depend on atomic motions, comparisons of Mössbauer and X-ray data have been used to estimate a disorder contribution for the iron atom; the value obtained is $\langle\Delta r_{Fe}^2\rangle \simeq 0.08$ Å^2 (this chapter, Sect. G). Although this value is only approximate, it nevertheless indicates that the observed values for $3B/8\pi^2$ (e.g., on the order of 0.44 Å^2 for backbone and 0.50 Å^2 for sidechain atoms) are dominated by the motional contribution. Thus, most experimental B-factor values are compared directly with the molecular dynamics results (i.e., neglecting the disorder contribution) or are rescaled by a constant amount (e.g., by setting the smallest observed B factor to zero) on the assumption that the disorder contribution is the same for all atoms.[37]

In Fig. 13 we show the calculated values of $\langle\Delta r_j^2\rangle$ obtained for the polypeptide backbone atoms of hen egg white lysozyme in a 100-ps simulation by averaging over two 50-ps segments.[192] For most of the backbone, the atomic fluctuations have converged in this period. The calculations are compared with the values estimated from the X-ray B factors without any disorder correction. It can be seen that in general there is good agreement between the simulation and experimental estimates, although some of the maxima in the two curves are displaced relative to each other by one or two residues. Another way of comparing simulation and X-ray results is shown in Fig. 60, which presents a correlation diagram of the fluctuations for lysozyme.[191] The approximate agreement between the theoretical and experimental estimates confirms the reliability of both the simulation and the temperature factors.

The great potential of the X-ray data for obtaining motional information has recently led to a molecular dynamics test[197] of the standard refinement techniques that assume isotropic and harmonic motion. Since simulations have shown that the atomic fluctuations are highly anisotropic and, in some cases, anharmonic (see Chapt. VI.A.1), it is important to determine the errors introduced in the refinement process by their neglect. A direct experimental estimate of the errors resulting from the assumption of isotropic, harmonic temperature factors is difficult because sufficient data are not yet available for protein crystals. Moreover, any data set includes other errors that would obscure the analysis, and the specific correlation of temperature factors and motion is complicated by the need to account for static disorder in the crystal. As an alternative to an experimental analysis of the errors in the refinement of proteins, a purely theoretical approach has been used.[197] The basic idea is to generate "X-ray" data from a molecular dynamics simulation

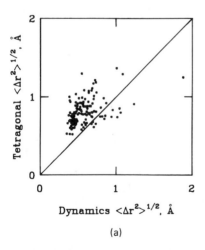
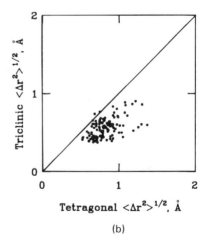

(a)   (b)

Figure 60. Correlation diagram for backbone rms fluctuations by residues: (a) tetragonal crystal versus simulation; (b) tetragonal crystal versus trigonal crystal.

of a protein and to refine these data by a standard procedure. The error in the analysis is determined by comparing the refined "X-ray" structure and temperature factors with the average structure and the mean-square fluctuations from the simulation. Such a comparison, in which no real experimental results are used, avoids problems due to inaccuracies in the measured data (exact calculated intensities are used), to crystal disorder (there is none in the model), and to approximations in the simulation (the simulation is exact for this case). The only question about such a test is whether the atomic motions found in the simulation are a meaningful representation of those occurring in proteins. As we have shown, comparisons suggest that molecular dynamics simulations provide a reasonable picture of the motions in spite of errors in the potentials, the neglect of the crystal environment and the finite time classical trajectories used to obtain the results. However, as already stated, these inaccuracies do not affect the exactitude of the computer "experiment" for testing the refinement procedure that is described here.

A 25-ps molecular dynamics trajectory for myoglobin[190] was used to generate the data. The average structure and the mean-square fluctuations from that structure were calculated directly from the trajectory. To obtain the average electron density, appropriate atomic electron distributions were assigned to the individual atoms and the results for each coordinate set were averaged over the trajectory. Given the symmetry, unit cell dimensions, and position of the myoglobin molecule in the unit cell, average structure factors, $\langle F(Q) \rangle$, and intensities, $I(Q) = |\langle F(Q) \rangle|^2$, were calculated as the Fourier transform of the average electron density, $\langle \rho(r) \rangle$, as a function of position r in the unit cell. Data were generated at 1.5-Å resolution, as this is comparable to the resolution of the best X-ray data currently available for proteins the size of myoglobin.[414,415] The resulting intensities at Bragg reciprocal lattice points were used as input data for the widely applied crystallographic program, PROLSQ.[416] The time-averaged atomic positions obtained from the simulation and a uniform temperature factor provide the initial model for refinement. The positions and an isotropic, harmonic temperature factor for each atom were then refined iteratively against the computer-generated intensities in the standard way.

Differences between the refined results for the average atomic positions and their mean-square fluctuations and those obtained from the molecular dynamics trajectory are due to errors introduced by the refinement procedure. The overall rms error in atomic positions ranged from 0.24 to 0.29 Å for slightly different restrained and unrestrained refinement procedures.[197] The errors in backbone positions (0.10 to 0.20 Å) are less than those for sidechain atoms (0.28 to 0.33 Å rms). These backbone errors, though small, are comparable to the rms deviation of 0.21 Å between the positions of the backbone atoms in the refined experimental structures of oxymyoglobin and carboxymyoglobin.[414,415] The positional errors are not uniform over the whole structure. They are plotted as a function of residue number in Fig. 61. There is a strong correlation between positional error and the magnitude of the mean-square fluctuation for an atom, with certain regions of the protein, such as loops and external sidechains, having the largest errors in their refined positions.

The refined mean-square fluctuations are systematically smaller than the fluctuations calculated directly from the simulation. The magnitudes and variation of temperature factors along the backbone are very well reproduced by the refinement (Fig. 62a), but the refined sidechain fluctuations are almost always significantly smaller than the actual values (Fig. 62b). Regions of the protein that have high mobility have large errors in temperature factors as well as in positions. A scatter plot of the fluctuations for all atoms (Fig. 63) shows that fluctuations greater than about 0.75 Å^2 ($B = 20$ Å^2) are almost always underestimated by the refinement. This figure demonstrates that the B factors (mean-square fluctuations) obtained from the refinement have an effective upper limit independent of the actual values calculated from the dynamics. This arises from the fact that most of the atoms with large fluctuations have multiple conformations and that the refinement procedure picks out one of them.

Because of the limited resolution of X-ray data for proteins, refinements that take some account of anisotropic motions have introduced assumptions concerning the nature of the anisotropy. One possibility is to assume anisotropic rigid body motions for sidechains such as tryphophan and phenyl-

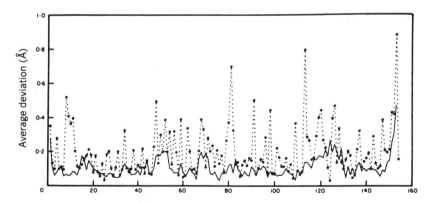

Figure 61. Positional error in refinement of the 25-ps data from myoglobin simulation: The deviations between atomic positions are calculated after the molecular dynamics average structure and the refined structures are superimposed by least squares. The deviations for backbone atoms (N, C, and C$^\alpha$, solid line) and sidechain atoms (dotted line) are averaged over residues.

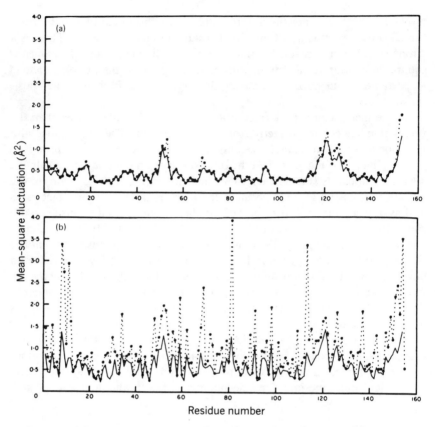

Figure 62. Residue averages of mean-square fluctuations from molecular dynamics (dotted line) and refinements (solid line). All plots are for the results of refining the 25-ps data from myoglobin simulation. (*a*) Backbone (N, C, and C$^\alpha$) averages; (*b*) sidechain averages.

alanine.[417,418] An alternative is to introduce a "dictionary" in which the orientation of the anisotropy tensor is related to the stereochemistry around each atom;[416] this reduces the six independent parameters of the anisotropic temperature factor tensor B_j to three parameters per atom. An analysis of a simulation for BPTI[419] has shown that the actual anisotropies of the atomic motions are generally not simply related to the local stereochemistry; an exception is the main-chain carbonyl oxygen, which has its largest motion perpendicular to the C=O bond. Thus, use of stereochemical assumptions in the refinement yields incorrectly oriented anisotropy tensors and reduced values for the anisotropies. Figure 64 shows ORTEP drawings of the thermal ellipsoid from the simulation and the refinement for Leu-29 of BPTI. Both the reduction in the effective anisotropy and the difference in orientation of

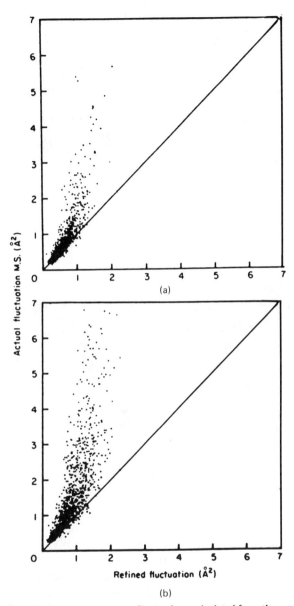

(a)

(b)

Figure 63. Scatter plots or mean-square fluctuations calculated from the myoglobin simulation and from the refinements. All the atoms are included in these plots. The exact mean-square fluctuations, $\langle \Delta r_j^2 \rangle$, calculated directly from the simulations, are plotted along the Y axis. The refined mean-square fluctuations, obtained from the refined temperature factors, are plotted along the X axis. (a) Results of a restrained refinement of the 25-ps data; (b) results from a refinement of the 300-ps simulation data with loose restraints.

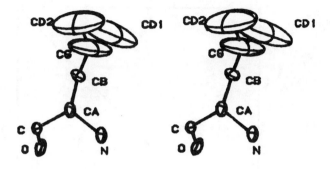

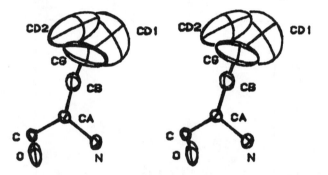

Figure 64. ORTEP stereo drawings for thermal ellipsoids of Leu-29 in BPTI in the dynamics (upper) and stereochemical (lower) principal-axis frames.

the anisotropy tensor are clear. As has been pointed out in Chapt. VI.A.1 and 2, the large-scale fluctuations of atoms have a collective character and side-chains tend to move as a unit so that the directions of largest motion are not related to the local bond direction; instead they have similar orientations for the different atoms that form a group which is undergoing correlated motions.

In a few cases very high resolution results (better than 1 Å resolution) have been obtained for small proteins, so that there are enough data for a full an-isotropic refinement; i.e., all six components of the atomic motional tensor, B_j, can be refined independently. One such system is the avian pancreatic polypeptide (aPP), a 36-amino acid hormone whose structure has been determined at 0.98-Å resolution.[417] A molecular dynamics simulation has been performed for aPP in a crystalline environment, treating the full unit cell (4 aPP molecules and 388 water molecules) for a time period (equilibration plus analysis) of 15 ps.[420] For some of the aromatic rings, the orientation of the calculated anisotropy tensor is similar to the experimental result. However, the calculated anisotropies are always larger than the experimental estimates. From the analysis of refinement procedures described above, this difference is likely to be due, at least in part, to errors in the refinement.

It has also been demonstrated that molecular dynamics can play a useful role in the refinement of protein structures against X-ray data.[420a] By adding an effective potential that represents the difference between the observed and calculated structure factors (Eq. 99) to the standard empirical potential function (Eq. 6), simulated annealing[420b] can be used to automatically refine a crude X-ray structure. In this way much of the manual rebuilding of the model structure, that is, the most time-consuming part of standard structure refinement,[421] can be avoided.

B. NUCLEAR MAGNETIC RESONANCE

Nuclear magnetic resonance (NMR) is an experimental technique that has played an essential role in the analysis of the internal motions of proteins.[29,30a,39a,32,33,421] Like X-ray diffraction, it can provide information about individual atoms; unlike X-ray diffraction, NMR is sensitive not only to the magnitude but also to the time scale of the motions. Nuclear relaxation processes are dependent on atomic motions on the nanosecond-to-picosecond time scale. Although molecular tumbling is generally the dominant relaxation mechanism for proteins in solution, internal motions contribute as well; for solids, the internal motions are of primary importance. In addition, NMR parameters such as nuclear spin-spin coupling constants and chemical shifts, depend on the protein environment. In many cases different local conformations exist but the interconversion is rapid on the NMR time scale of milliseconds, so that average values are observed. When the interconversion time is on the order of the NMR time scale or slower, the transition rates can be studied; an example is provided by the reorientation of aromatic rings (see Chapt. VI.B.1 and below).

The measurements of NMR relaxation parameters (T_1, T_2, and NOE) provide information concerning the time scales and the amplitudes of atomic motions. Under most circumstances, the relaxation behavior of the nuclear spins (^1H, ^{13}C) in a protein is governed by dipolar interactions between the various spins.[422,423] The dipolar couplings depend on the relative positions of the coupled nuclei and on their relative motions.[423,424] We consider spin $1/2$ nuclei in what follows since they are most important for protein NMR. For spin-lattice relaxation, such as observed in T_1 or nuclear Overhauser effect (NOE) measurements, it is possible to express the behavior of the magnetization in the form[425,426]

$$\frac{d(I_z(t) - I_0)_i}{dt} = -\rho_i(I_z(t)_i - I_{0i}) - \sum_{i \neq j} \sigma_{ij}(I_z(t)_j - I_{0j}) \quad (101)$$

where $I_z(t)_i$ and I_{0i} are the z components of the magnetization of nucleus i at time t and at equilibrium, respectively; ρ_i is the direct relaxation rate of nucleus i; and σ_{ij} is the cross-relaxation rate between nuclei i and j. Similar expressions can be derived for spin-spin relaxation such as can be observed in T_2 or line-width measurements. The quantities ρ_i and σ_{ij} can be expressed in terms of spectral densities, $J(\omega)$, by the equations

$$\rho_i = \frac{6\pi}{5} \gamma_i^2 \hbar^2 \sum_{j \neq i} \gamma_j^2 [\tfrac{1}{3} J_{ij}(\omega_i - \omega_j) + J_{ij}(\omega_i) + 2J_{ij}(\omega_i + \omega_j)] \quad (102)$$

$$\sigma_{ij} = \frac{6\pi}{5} \gamma_i^2 \gamma_j^2 \hbar^2 [2J_{ij}(\omega_i + \omega_j) - \tfrac{1}{3} J_{ij}(\omega_i - \omega_j)] \quad (103)$$

where ω_i is the Larmor frequency of nucleus i.

The spectral density functions can be obtained from the correlation functions for the relative motions of the nuclei with spins i and j;[423,425] they have the form

$$J_{ij}^n(\omega) = \int_0^\infty \left\langle \frac{Y_n^2[\theta_{lab}(t)\phi_{lab}(t)] \, Y_n^{2*}[\theta_{lab}(0)\phi_{lab}(0)]}{r_{ij}^3(0)r_{ij}^3(t)} \right\rangle \cos \omega t \, dt \quad (104)$$

where $Y_n^2[\theta(t)\phi(t)]$ are second-order spherical harmonics and the angular brackets represent an ensemble average, which is approximated by an integral over the molecular dynamics trajectory. The quantities $\theta_{lab}(t)$ and $\phi_{lab}(t)$ are the polar angles at time t of the internuclear vector between nucleus i and j with respect to the external magnetic field and r_{ij} is the internuclear distance. In the simplest case of a rigid molecule undergoing isotropic tumbling with a correlation time τ_0, Eq. 104 reduces to the familiar expression

$$J_{ij}(\omega) = \frac{1}{4\pi r_{ij}^6} \frac{\tau_0}{1 + (\omega\tau_0)^2} \quad (105)$$

The contribution of the internal motions to the spectral density is particularly simple to evaluate for T_1 values associated with carbon-13 nuclei that are singly protonated. Because ^1H spins are generally saturated in the experiment, the second term of Eq. 101 vanishes. Further, because the distance from the carbon to its directly bonded proton is short compared with the distance to other protons, the latter can be ignored and Eq. 103 reduces to that describing the motion of the single pair. As r_{ij} is essentially fixed for a $C-H$ bond, only the angular variation in Eq. 104 is important.[329,423,425] In the more general cases where several nonbonded spin pairs are involved and where the r_{ij} values are time dependent, the full analysis, including averaging over a number of interactions, is required; details of these procedures are given elsewhere.[425,427]

The internal motion correlation functions can be evaluated directly from molecular dynamics simulations. Hence the relaxation behavior described by Eq. 101 can be obtained from such simulations. Figure 65 presents a schematic diagram of the type of correlation function that is associated with the internal motions. The rapid loss of correlation in the first few picoseconds results from the librational motion of the residue containing the nucleus and from collisions between the atoms of the residue and those of the surrounding protein cage. An intermediate plateau value is often reached, followed by a slow loss of correlation over several hundred picoseconds or longer. The latter arises from larger-scale and more complex motions, including rare events, such as dihedral angle transitions between minima separated by high barriers (see Chapt. VI.B.1).

Figure 66 shows correlation functions for several protonated carbons of BPTI obtained from a molecular dynamics simulation.[329] In these, as in most cases, the motional averaging over the time scale of the simulation is largely

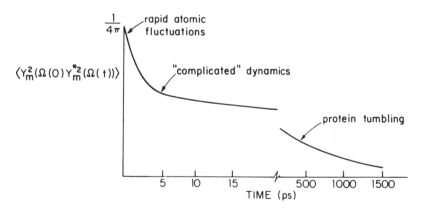

Figure 65. Schematic diagram of the general behavior of the decay of NMR correlation functions in proteins. The physical origins for the decay in correlation resulting from the different kinds of protein motions are indicated.

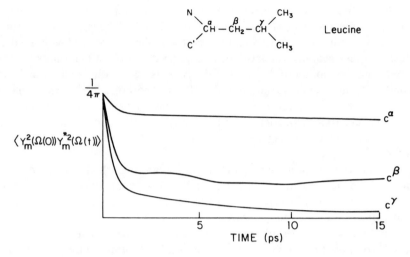

Figure 66. NMR correlation functions for C^α, C^β, and C^γ of Leu-29 calculated from a BPTI molecular dynamics simulation.

complete after 2 ps. This time is very much less than the NMR resonance frequencies, so the effect is simply to reduce the relaxation rates from those expected for a rigidly tumbling molecule by a factor depending on the extent of motional averaging; i.e., it is possible to write an expression for T_1 of the form

$$(1/T_1)_{\text{actual}} = S^2(1/T_1)_{\text{rigid}} \tag{106}$$

where $(1/T_1)_{\text{actual}}$ is the relaxation rate with motional averaging, $(1/T_1)_{\text{rigid}}$ is that obtained for the rigidly tumbling molecule (Eq. 105), and S^2 is an order parameter ($0 \le S^2 \le 1$) that is related to the magnitude of the fast motions involved.[329,333,428] The resulting order parameters for ^{13}C nuclei obtained from molecular dynamics simulations, relative to those from the rigid model, are shown in Table XI.[329] For the α-carbons of BPTI, the motional averaging effects cause a decrease in $(1/T_1)$ of less than 20% in nearly all the cases examined; this is in accord with the limited experimental data available. As one moves out along the sidechains, much smaller-order parameters are predicted, up to a factor of 0.25 for the motions contributing to the 2-ps decay. Further, as Fig. 66 shows, the importance of events on a longer time scale than 2 ps increases, reflecting the greater contribution of lower-frequency fluctuations to sidechain motions than to those of the protein backbone. These slower motions are likely to require more extensive calculations to obtain quantitatively reliable predictions for T_1 values. Nevertheless, the

TABLE XI
Motional Averaging Order Parameter for ^{13}C NMR
Spin-Lattice Relaxation Times in BPTI[a]

Residue	Motional Averaging Scale Factor[b]	
	2 ps	96 ps
α-Carbons		
Phe-22	0.91	0.89
Gly-28	0.84	0.84
Leu-29	0.91	0.89
β-Carbons		
Leu-6	0.65	0.29
Ile-18	0.76	0.37
Ile-19	0.89	0.88
Leu-29	0.39	0.31
γ-Carbons		
Leu-6	0.54	0.25
Ile-18	0.56	0.27
Ile-19	0.51	0.51
Leu-29	0.24	0.09

[a] Obtained from molecular dynamic simulation; see Ref. 329.
[b] Values of correlation functions at 2 ps and 96 ps were used as indicated to determine the order parameters.

available relaxation data correspond well to the extent of motional averaging obtained from the simulations. In BPTI, for example, there is evidence that greater increases in T_1 values occur for sidechain atoms, e.g., isoleucine C^δ, than for C^α atoms.[429] Similar behavior has been found for the rather more complex cases of nonprotonated carbons that have been studied.[427] It should be noted that the calculated relaxation times would be in substantial disagreement with experiment if, as has been suggested recently,[385] the motions occurred on a nanosecond time scale instead of the picosecond time scale found in the dynamics simulations.

A comparison of experimental order parameters with those calculated from a molecular dynamics simulation has been made for 12 of the methyl groups of BPTI.[430] The relative flexibility of the residues studied is well described by the simulation, although the theoretical order parameters indicate somewhat less motion than the experimental results. Similar conclusions have been reached in an analysis of the sidechain order parameters for the leucines in myoglobin calculated from a 300-ps simulation.[190]

The nuclear Overhauser effect (NOE) corresponds to the selective enhancement of a given resonance by the irradiation of another resonance in a dipolar coupled spin system. Of particular interest for obtaining motional information are measurements that provide time-dependent NOEs from which the cross-relaxation rates σ_{ij} (see Eq. 103) can be determined directly or indirectly by solving a set of coupled equations (Eqs. 101 to 103). Motions on the picosecond time scale are again expected to introduce averaging effects that decrease the cross relaxation rates by a scale factor relative to the rigid model. A lysozyme molecular dynamics simulation[191] has been used to calculate dipole vector correlation functions[425] for proton pairs that have been studied experimentally.[431] Four proton pairs on three sidechains (Trp-28, Ile-98, and Met-105) with very different motional properties were examined. Trp-28 is quite rigid, Ile-98 has significant fluctuations, and Met-105 is particularly mobile, in that it jumps among different sidechain conformations during the simulation. The results for the order parameters obtained from the simulation and experiment are shown in Table XII. The ranking of the order parameters is the same in the theoretical and experimental results. However, although the results for the Trp-28 protons agree with the measurements to within the experimental error, for both Ile-98 and Met-105 the motional averaging found from the NOEs is significantly greater than the calculated values. This suggests, in accord with the ^{13}C relaxation results, that there are rare fluctuations involving transitions that are not adequately sampled by the 40 ps simulation.

One way to obtain dynamic results for longer time periods is to use stochastic dynamics methods (see Chapt. IV.D). As described in Chapt. IX.B.3, these have been used to simulate aliphatic sidechains in aqueous solution for periods of 100 ns or longer so that they permit adequate sampling of both the oscillations within a potential well and the transitions between wells.[338] To provide an overview of the type of ^{13}C NMR relaxation behavior that can

TABLE XII
Motional Averaging Order Parameter for Proton Cross-Relaxation
Rates in Lysozyme

Residue	Vector	Motional Averaging Scale Factor	
		Dynamics	Experimental
Trp-28	$H^{\epsilon 3}$-$H^{\zeta 3}$	0.86	0.91 ± 0.10
Trp-28	$H^{\zeta 3}$-$H^{\eta 3}$	0.83	0.93 ± 0.07
Ile-98	$H^{\gamma 11}$-$H^{\gamma 12}$	0.82	0.63 ± 0.08
Met-105	$H^{\beta 1}$-$H^{\beta 2}$	0.69	0.49 ± 0.05

Source: Ref. 191.

result from such motions,[423] a sidechain with four internal degrees of freedom was used and the tumbling time τ_0 of the protein to which it was attached was varied over the range 1 to 100 ns; this includes the rotational diffusion behavior of a range of proteins at different concentrations in a variety of solvents (e.g., dilute aqueous solutions of BPTI and lysozyme have tumbling times on the order of 1 ns and 10 ns, respectively, at room temperature). Also, the dependence of the motion on the spectrometer frequency was examined by doing the relaxation parameter calculations at two frequencies (15 MHz and 65 MHz) appropriate for ^{13}C measurements. The results obtained are shown in Table XIII, which gives the calculated T_1, T_2, and NOE values; the carbon atoms are labeled such that C_2 is tumbling rigidly with the macromolecule (i.e., its correlation function is given by Eq. 105), while the other carbons (C_3, C_4, C_5, and C_6 have an increasing number of dihedral-angle degrees of freedom (one for C_3 to four for C_6) intervening between the protein and the ^{13}C nucleus under study. It is evident from the table that there are a variety of

TABLE XIII
Theoretical Sidechain NMR Parameters

	15 MHz			68 MHz		
	T_1 (ms)	T_2 (ms)	NOE $(1 + \eta)$	T_1 (ms)	T_2 (ms)	NOE $(1 + \eta)$
			$\tau_0 = 1$ ns			
C_2 (analytic)	26.5	25.8	2.80	58.3	48.9	1.70
C_3	112	110	2.86	179	163	2.34
C_4	290	290	2.97	312	308	2.90
C_5	452	450	2.96	493	486	2.90
C_6	627	625	2.97	682	672	2.90
			$\tau_0 = 10$ ns			
C_2 (analytic)	11.9	7.3	1.30	133	11	1.16
C_3	60	40	1.63	230	57	2.44
C_4	243	223	2.74	292	247	2.91
C_5	368	313	2.50	492	358	2.95
C_6	519	454	2.60	665	515	2.92
			$\tau_0 = 100$ ns			
C_2 (analytic)	62.4	1.2	1.16	1,270	1.2	1.15
C_3	171	7.1	2.18	302	7.2	2.87
C_4	274	101	2.93	293	103	2.93
C_5	466	99	2.86	496	100	2.94
C_6	617	163	2.89	667	165	2.94

Source: Ref. 423.

motional effects on the relaxation parameters. Important variations in the results are found for different tumbling times and spectrometer frequencies. Particularly striking is the fact that at short and intermediate tumbling times ($\tau_0 = 1$ ns, 10 ns), the values of T_1 increase with greater motional freedom, while they decrease at the longest tumbling time ($\tau_0 = 100$ ns). By contrast, T_2 increases with greater motional freedom for all tumbling times. This makes clear the importance of measuring both T_1 and T_2, ideally at different spectrometer frequencies, to obtain as complete an experimental data set as possible.

Some comparisons were made of the full stochastic simulation results with simplified models that are often used in analyzing NMR data. Two models were considered. The first is the product approximation, in which it is assumed that the motions about the different dihedral angles are uncorrelated. In general, the true correlated motions lead to less averaging than those given by the product approximation, which therefore generally results in longer relaxation times than those from the complete calculation. The second model is the lattice jump approximation, which has been widely used for dihedral-angle transitions.[432,433] Very good agreement with the full calculations was obtained with the lattice jump model in the present case. However, this was shown to be due to a cancellation of errors in the lattice jump model, involving the neglect of relaxation by small oscillations within a well and the neglect of coupling between the transitions of more than one dihedral angle. Thus, although the lattice jump model works well here, it might not at lower temperatures or in the protein interior, where correlated motions and small oscillations are of greater importance.

The experimental studies of internal motions considered so far involve essentially fixed distance interactions (i.e., $^{13}C-H$ interactions or protons at fixed distances). If nuclear Overhauser effects are measured between protons whose distances are not fixed by the structure of a residue, the strong distance dependence of the cross-relaxation rates ($1/r^6$) can be used to obtain estimates of the interproton distances.[431,434] The simplest application of this approach assumes that proteins are rigid and tumble isotropically. To determine whether picosecond fluctuations are likely to introduce important errors into the interproton distances obtained by this procedure, a molecular dynamics simulation was analyzed.[425] Simulations for lysozyme suggest that the presence of the motions will cause a general decrease in most NOE effects observed in a protein. The decrease is too small, however, to produce a significant change in the distances estimated from the measured NOE values. This is consistent with the excellent correlation found between experimental NOE values and those calculated by using distances from a crystal structure.[431] Specific NOEs can, however, be altered by the internal motions to such a degree that the effective distances are considerably different from those predicted for a static structure. Such possibilities must, therefore, be considered in any structure determination based on NOE data. This is true particularly for cases involving averaging over large fluctuations.

NOE values for approximate interproton distances in macromolecules have been employed in examining the feasibility of determining the solution structure of a protein from the distance constraints available from such measurements. Although distance geometry algorithms have been found useful,[435] the fact that the available constraints from NOE data are approximate and limited to distances of less than 5 Å suggests that introduction of additional structural or energetic information may be important for obtaining reliable results. Molecular dynamics simulations were performed with an effective potential function that consists of the standard empirical energy expression (Eq. 6) augmented by biharmonic terms to represent the NOE restraints.[436,437] In a model study of the small protein crambin,[438] it was found that with realistic NOE restraints, the molecular dynamics simulations converged to the known crambin structure from different extended structures; the average structure obtained from the simulations had rms deviations of 1.3 Å and 1.9 Å for the mainchain and sidechains atoms, respectively. Furthermore, it was shown that average dynamics structures with significantly larger deviations could be characterized as incorrect, independent of a knowledge of the crystal structure. Clearly, the available NOE folding studies are still at an exploratory stage but they do raise the hope that structure determination by NMR in solution will be a feasible supplement to X-ray diffraction studies of crystals. It has been demonstrated, in fact, that the NMR solution structure can be used as a starting point for the molecular replacement method in refining X-ray data.[439]

In Chapt. VI.B.1 tyrosine ring flips in BPTI were considered as models of activated processes in proteins. One reason for choosing the ring-flipping "reaction" was the fact that NMR data on the rates of aromatic ring rotations were available. Although the flipping rate is often too fast to be measured directly (> 2000 s^{-1}), specific ring flipping rates have been studied by varying the temperature and following the coalescence of the resonances for the pairs of ortho and meta ring protons[119,440] or by cross-saturation measurements.[440] For Tyr-35 in BPTI, the ring rotation rate has been measured to be ≈ 0.6 s^{-1} at 300 K.[119] For temperatures between 40 and 80°C, analysis of the rates, assuming that the enthalpy and entropy of activation are independent of temperature, yielded $\Delta H^\dagger \simeq 37$ kcal/mol and $\Delta S^\dagger \simeq 68$ entropy units. These results differ from the calculated results for Tyr-35 (Chapt. VI.B.1) in two ways. First, the experimental rate constant at 300 K is much smaller than the calculated value. Second, there is a large positive entropy contribution to the free energy of activation, in contrast to the small calculated value. If the preexponential factor calculated from the simulation is correct as to the order of magnitude, the difference between the theoretical and experimental rate constant must arise from a difference in the effective barrier; as pointed out in Chapt. VI.B.1, a value of 17 kcal/mol instead of that calculated from the umbrella sampling analysis (9.8 kcal/mol) would be required. For a system as large as a protein, such a difference would not be surprising, particularly when one considers that the dominant contribution to the energy barrier comes from the surrounding matrix atoms rather than from intrinsic ring torsional terms. It is of interest in this regard that the adiabatic barrier obtained from the dynamically equilibrated structure (11 kcal/mol) is 5 kcal/mol lower than that (16 kcal/mol) from the X-ray structure. Thus equilibration of the activated ring configurations appears to have produced a lower barrier as a result of unphysical relaxation of the surrounding protein matrix. A possible source of this relaxation effect is the absence of solvent surrounding the protein; for example, in molecular dynamics simulations in the presence of solvent, the average dynamics structure remained closer to the X-ray than that obtained from a vacuum simulation (see Chapt. IX.A).[193,441] Although Tyr-35 is a relatively buried residue, it is in fact situated close to the protein surface, as are almost all residues in BPTI because it is such a small protein. This suggests that the rotation of Tyr-35 is accompanied by distortions of the protein surface and thus would be sensitive to the presence of solvent. A simulation[442] using a 10-Å reaction zone with solvent included yielded a higher barrier (13 kcal/mol) than the earlier activated dynamic study (Chapt. VI.B.1) and the preexponential factor remained essentially unchanged. The calculated rate (5.4×10^2 s^{-1}) is still in significant disagreement with experiment, suggesting that other refinements (e.g., a potential function including the ring hydrogens explicitly) are required.

As to the large activation entropy and the concomitantly large activation enthalpy found in the experimental analysis of the temperature dependence of the reaction,[119] a number of considerations suggest that these values need to be interpreted with care. First, as already discussed in Chapt. VI.B.1, use of the standard transition-state formulation with a quantum-mechanical preexponential factor ($k_B T/h$) is not valid for a solution reaction; an extension of the classical Kramers formulation to the intermediate-viscosity regime is more appropriate. However, by coincidence, the preexponential factor obtained from the activated dynamics simulations, $\frac{1}{2}\kappa \langle |\dot{\xi}| \rangle = 2.2 \times 10^{12}$ s^{-1} at 300 K, is very close to the value of ($k_B T/h$) (6×10^{12} s^{-1}). More important is the fact that the experimental quantities ΔH† (which is essentially ΔE† because the pV term is expected to be negligible) and ΔS† include contributions from a variety of factors, in addition to the mean energy and entropy difference, respectively, between the initial and activated states at a given temperature. Because the protein matrix contributes a significant part of the barrier, the ΔE† value is expected to have a temperature dependence.[443] The thermal expansion of the protein matrix would lead to a negative, temperature-dependent contribution to $\Delta E\dagger$; over the temperature range studied, this would result in an increase in the apparent $\Delta E\dagger$ and $\Delta S\dagger$. Additional contributions due to the temperature dependence of the solvent viscosity and internal viscosity are also expected. In an experimental analysis of the rebinding of oxygen in myoglobin,[318] it was shown that the intrinsic energy and entropy of activation of the reaction are much reduced if the effect of viscosity and its temperature dependence are separated out by isoviscosity measurements.

It is also possible that the separation of the free energy into enthalpy and entropy terms in the calculation is inaccurate. If the reaction coordinate ξ used in the analysis were not an adequate reaction coordinate, most of the protein configurations with $\xi = \xi\dagger$ would be located below the transition-state ridge, resulting in a larger calculated value of $\rho(\xi\dagger)$. Also, a smaller fraction of the trajectories initiated from configurations with $\xi = \xi\dagger$ would lead to successful attempts to cross the barrier: this would lead to a smaller value of κ. Although the net result for the overall rate constant k and free energy of activation $\Delta G\dagger$ would be unchanged,[127] there would be an apparent negative entropy contribution, making the calculated entropy smaller than the experimental value, as observed. However, because the calculated transmission coefficient is near unity, the effect is expected to be small relative to the apparent discrepancy. To correctly determine the values of ΔH^+ and $\Delta S\dagger$, the rate constant would have to be calculated for a series of temperatures and the result fitted to an Arrhenius-type expression to separate the temperature-dependent and temperature-independent contributions.

Also of interest is the effect of hydrostatic pressure on reactions in the protein interior. An NMR study has been made of the pressure dependence of the aromatic ring rotations in BPTI.[444] The experimental activation volume, $\Delta V\dagger$, determined from the pressure dependence of the rate constant by the equation

$$\frac{\partial \ln k}{\partial P} = -\frac{\Delta V\dagger}{k_B T} \tag{107}$$

was found to be about 50 Å^3, with the positive sign for $\Delta V\dagger$ corresponding to a decrease in rate with increasing pressure. The observed magnitude of the activation volume, on the order of that associated with protein denaturation, provides an important test for the theoretical interpretation of the ring rotation process. For motions in the interior of a protein,[445] as for solution reactions[131a] in general, the pressure dependence of the rate constant is not related directly to a physical volume change between the reactant and transition state. Instead, it is expected to be dominated by the interactions between the reacting species and the solvent environment, which in the case of the tyrosine ring rotations is provided by the surrounding protein atoms.[130] To analyze the factors involved,[446] the Kramers formulation (Eqs. 82 and 83) for an activated process in the diffusive limit was used. It was shown that the primary effect on the rate constant arises from the decrease in the distance between the ring and the surrounding protein atoms when the protein is compressed. Both the activation energy and the friction coefficient (internal viscosity) are calculated to increase with increasing pressure. An estimate of the magnitudes of the two effects indicates that both are important and give activation volumes on the order of the experimental value. Thus, rather than a large physical volume change, what is measured is the pressure dependence of rather small packing defects, which have been shown to play an essential role in initiating the ring rotation (see Chapt. VI.B.1).[446]

NMR spin-spin coupling constants and chemical shifts also can be analyzed by use of dynamical simulations of proteins. Because of the NMR time scale, the measured values of these parameters represent averages over the internal fluctuations. To obtain an understanding of the role of fluctuations two different ways of determining the averages from the simulations can be compared. The first procedure makes use of the appropriate formulas (e.g., expressions for the ring current contribution to the chemical shift[447,448] or the dihedral angle dependence of the vicinal coupling constant[449]) and calculates the distances and angles required from the average structure obtained from the simulation. This corresponds to what is done to estimate the NMR parameter from an X-ray structure.[450] However, such a calculation is not correct for a fluctuating molecule. Instead, one should calculate the value of the chemical shift or coupling constant for each dynamics coordinate set and then average the physical quantity over the trajectory. The differences between the two procedures determines the effects of fluctuations on the values of the NMR parameters. For ring-current-induced chemical shifts in BPTI, the simulation results yielded fluctuations up to ± 6 ppm, on the order of the average chemical shift values themselves.[451] However, the distribution of the fluctuations is such that the proper average over the dynamics and the calculation based on the average structure generally result in very similar values. Vicinal spin-spin coupling constants show a different behavior.[452] The differences between the two types of averages vary from zero to more than 6 Hz; the values of the coupling constants themselves are between 2 and 14 Hz. Examination of the dynamics demonstrates that all cases where large differences occur correspond to dihedral angle fluctuations with jumps between different minima. The differences are larger for sidechain than for mainchain angles and increase as one goes outward along a sidechain. Analysis of the simulation results indicates that under appropriate conditions, coupling constant measurements can be used to determine the magnitude of the torsion angle fluctuations.[452]

C. FLUORESCENCE DEPOLARIZATION

Fluorescence depolarization measurements of aromatic residues and other probes in proteins can provide information on the amplitudes and time scales of motions in the picosecond-to-nanosecond range. As for NMR relaxation, the parameters of interest are related to time correlation functions whose decay is determined by reorientation of certain vectors associated with the probe (i.e., vectors between nuclei for NMR relaxation and transition moment vectors for fluorescence depolarization). Because the contributions of the various types of motions to the NMR relaxation rates depend on the Fourier transform of the appropriate correlation functions, it is difficult to obtain a unique result from the measurements. As described above, most experimental estimates of the time scales and magnitudes of the motions generally depend on the particular choice of model used for their interpretation. Fluorescence depolarization, although more limited in the sense that only a few protein residues (i.e., tryptophans and tyrosines) can be studied with present techniques, has the distinct advantage that the measured quantity is *directly* related to the decay of the correlation function.

The fluorescence emission anisotropy, $r(t)$, at time t after excitation of the chromophore is defined by the expression

$$r(t) = \frac{I_\parallel(t) - I_\perp(t)}{I_\parallel(t) + 2I_\perp(t)} \tag{108}$$

where $I_\parallel(t)$ and $I_\perp(t)$ are the fluorescence intensities polarized parallel and perpendicular, respectively, to the polarization of the incidence beam at time t. For protein molecules in solution (i.e., with an isotropic initial orientation of the chromophore), Eq. 108 reduces to a simple correlation function expression[453,454]

$$r(t) = \tfrac{2}{5} \langle P_2[\hat{\mu}_A'(0) \cdot \hat{\mu}_E'(t)] \rangle \tag{109}$$

The quantities $\hat{\mu}_A'(t)$ and $\hat{\mu}_E'(t)$ are unit vectors in a space-fixed coordinate system directed along the chromophore absorption and emission dipole moments, respectively, $P_2(x)$ is the second-order Legendre polynomial, and the angular brackets denote an ensemble average.

When Eq. 109 is applied to a protein molecule which is much larger than the fluorescent probe (e.g., a tryptophan residue), the fast oscillations of the probe can be uncoupled from the much slower rotational motion of the protein. If the overall rotation is diffusional and isotropic, the anisotropy is given by[324,454a]

$$r(t) = \tfrac{2}{5} e^{-t/\tau_0} \langle P_2[\hat{\mu}_A(0) \cdot \hat{\mu}_E(t)] \rangle \tag{110}$$

where τ_0 is the correlation time for the overall rotation of the protein and $\hat{\mu}_A(0)$ and $\hat{\mu}_E(t)$ are the unit transition dipole vectors with components defined in a local coordinate system fixed in the molecule. For proteins the rotational correlation times are 1 ns or longer, so that for the picosecond dynamics studied in simulations, the separability of internal and overall motions assumed in Eq. 110 is valid.[423,428] If internal motions on a longer time scale are being considered, the simple form of Eq. 110 may no longer be applicable.[324,454a]

For a residue in a protein the range of internal motions is generally restricted, so that $\langle P_2[\hat{\mu}_A(0) \cdot \hat{\mu}_E(t)] \rangle$ is not expected to decay to zero. The correlation function usually decays, possibly with oscillations, to a plateau value on a picosecond time scale; this is analogous to the NMR results described above. A realistic case corresponds to an internal correlation function which separates into two time scales ($\tau_1 \ll \tau_2 \ll \tau_0$), for which we can write

$$r(t) = (r_0)_{\text{eff}}[(1 - P_\infty^2)e^{-t/\tau_2} + P_\infty^2]e^{-t/\tau_0} \tag{111}$$

and

$$(r_0)_{\text{eff}} = P_\infty^1 r_0 = S^2 \tag{112}$$

with S^2 ($0 \le S^2 \le 1$), the generalized order parameter (on a picosecond timescale). The quantities P_∞^1 and P_∞^2 are the plateau values associated with the τ_1 and τ_2 relaxation and it is assumed that τ_1 is short compared to the time resolution of the experiment. Equation 111 corresponds to the expression often used in the analysis of depolarization measurements.[455] The portion of the decay that is studied by the molecular dynamics simulation yields $(r_0)_{\text{eff}}$. The longer time decay due to the internal motions corresponding to the relaxation time τ_2 is observable experimentally (e.g., 100 ps $\le \tau_2 \le$ 1 ns); examples are given in a number of studies.[455,456]

In an analysis of the correlation function P_2 (see Eq. 110) for the internal motions of the four tyrosines in the bovine pancreatic trypsin inhibitor in a van der Waals solvent,[193] it was found that for three out of four tyrosines the correlation function P_2 decays from its initial value of unity to a plateau value of ~0.8 in less than 2 ps and then remains essentially constant for the rest of the simulation; for the fourth tyrosine (Tyr 10), which is on the surface of the molecule, the P_2 value continued to decay over the entire run. Consequently, the measured value of $(r_0)_{eff}$ on a nanosecond time scale should be significantly reduced relative to the ideal value ($r_0 = 0.4$) expected when the absorption and emission dipoles coincide. The importance of $(r_0)_{eff}$ was emphasized in an analogous but more detailed analysis[457] of the same tyrosines based on a vacuum simulation of BPTI;[207] results for $(r_0)_{eff}$ similar to those calculated from the solvent run were found. Stationary-state fluorescence polarization measurements of the tyrosines in BPTI that are consistent with such a reduced value of $(r_0)_{eff}$ have been reported;[458] the average value of the ratio $(r_0)_{eff}/r_0$ estimated from the tyrosine data is 0.78, although the interpretation in terms of tyrosine motion is not unequivocal due to the possibility of energy transfer.

Most experimental fluorescence depolarization studies of proteins have been concerned with tryptophans, rather than tyrosines.[455,456,459] To obtain information for interpreting these results, an analysis of the fluorescence depolarization of tryptophans in a molecular dynamics simulation of lysozyme was made.[324] Lysozyme is particularly useful for theoretical analysis because it contains six tryptophans in a variety of environments. Thus the range of the calculated motional behavior should reflect that occurring in most proteins. Because the detailed photophysics of tryptophan is not completely understood, a variety of models for absorption and fluorescence emission were analyzed. Some results for the simplest case, in which the absorption and emission dipole moment are assumed to have the same orientation, are shown in Fig. 67. For a lower limit of the time resolution greater than 20 ps, the calculated order parameters range from 0.94 for a completely buried tryptophan (Trp-108) to 0.36 for Trp-62 which is exposed in the active site. Because of the high exposure of Trp-62, the vacuum simulation results may be artificial (see Chapt. IX.B.2). Studies of tryptophans in a variety of proteins have led to values of $(r_0)_{eff}$ in the range 0.18 to 0.30.[455,456,459] Translation of these results into order parameters S^2 is difficult because the r_0 value (see Eq. 112) for tryptophan appears to be somewhat less than the ideal theoretical value of 0.4; i.e., even in propylene glycol at 58 K, r_0 values less than 0.4 are obtained. Nevertheless, the available experimental results suffice to indicate that subnanosecond relaxation of tryptophans occurs in proteins.

It is of interest to examine the nature of the motions of a sidechain as large as a tryptophan residue in the protein interior. Reorientation of the transition dipole vector of tryptophan is due mainly to rotation about the C_α—C_β and C_β—C_γ bonds, which correspond to the χ^1 and χ^2 dihedral angles of the sidechain, although larger-scale collective motions of the backbone are also involved. In the protein environment, the fluctuations of χ^1 and χ^2 are expected to be anticorrelated so that large variations in the two angles result in a

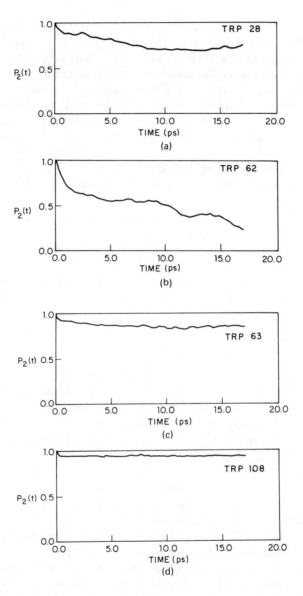

Figure 67. Correlation functions for fluorescence depolarization: (*a*) Trp-28; (*b*) Trp-62; (*c*) Trp-63; (*d*) Trp-108; (*e*) Trp-111; (*f*) Trp-123.

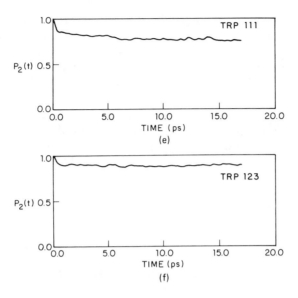

smaller net motion of the bulky ring system. The degree of correlation can be determined from the equal-time cross correlation function

$$C(\chi^1, \chi^2) = \frac{\langle \Delta\chi^1 \Delta\chi^2 \rangle}{\langle (\Delta\chi^1)^2 \rangle^{1/2} \langle (\Delta\chi^2)^2 \rangle^{1/2}} \tag{113}$$

where $\Delta\chi^j$ is the deviation of the angle χ^j from its mean value. The results for the tryptophans of lysozyme are shown in Table XIV. It is evident that for large fluctuations and small anticorrelation values, the P_2 motional averaging is greatest. Comparing Trp-28 and Trp-108, we see that the angular fluctuations are similar, but that for Trp-28, with the larger decay in P_2, the anticor-

TABLE XIV
Equal-Time Correlation for χ^1 and χ^2

Residue	$\langle (\Delta\chi^1)^2 \rangle^{1/2}$ (deg)	$\langle (\Delta\chi^2)^2 \rangle^{1/2}$ (deg)	$\dfrac{\langle \Delta\chi^1\Delta\chi^2 \rangle}{\langle (\Delta\chi^1)^2 \rangle^{1/2}\langle (\Delta\chi^2)^2 \rangle^{1/2}}$
28	10.2	19.4	−0.20
62	35.5	26.4	−0.61
63	11.9	15.3	−0.27
108	12.3	15.2	−0.78
111	10.5	21.9	−0.46
123	11.5	12.9	−0.24

Source: Ref. 324.

relation value is much less than that for Trp-108. For Trp-62 with the greatest decay, the anticorrelation is rather high but the individual angular fluctuations are considerably larger than for the other tryptophans. These results make clear that a model which treats each of the dihedral angles independently would not be valid for sidechain motions in the interior of a protein. This contrasts with the behavior of simple alkanes and exposed protein sidechains where an uncorrelated model gives satisfactory results for dipolar relaxation of ^{13}C nuclei in NMR (see this chapter, Sect. B, above).[423]

The study of fluorescence depolarization of tryptophans in proteins is likely to develop into a powerful tool for analyzing the internal motions.[459a] Clarification of the photophysics of tryptophans by jet measurements,[460] extension of the time scale of observation into the picosecond and femtosecond range,[461] and utilization of normal or modified proteins with only a single tryptophan[462] should lead to a significant increase in the utility of this experimental approach.

D. VIBRATIONAL SPECTROSCOPY

As described in Chapt. VI.A.3, harmonic calculations have shown that the vibrations of proteins span a range from 3 to 3000 cm^{-1}, corresponding to periods of 10 ps to 10 fs. In principle the vibrations can be probed by a variety of different spectroscopic techniques, including infrared absorption, resonant and nonresonant Raman scattering, and inelastic neutron scattering. Both infrared and Raman spectroscopy have been used to study the vibrations of polypeptides; e.g., detailed analyses have been made of polyalanine in both the α-helical and β-sheet configurations.[463-466] For proteins, fewer results are available other than for relatively high frequency motions associated with the localized bonding structure.[467,468] Small shifts in the frequencies of some of these vibrations (e.g., disulfide bonds) are observed going from the native to the denatured protein.[469] For chromophores, such as retinal in rhodopsin[470] and the heme group in myglobin and hemoglobin,[471,472] resonance Raman spectroscopy has been particularly useful in probing the nature and dynamics of conformational change.

Low-frequency vibrations (20 to 100 cm^{-1}) have been detected by Raman and infrared spectroscopy in several proteins.[36,259,473,474] However, the exact nature of these modes is not known, although the frequency range involved suggests that the motions have a large-scale, collective character. As mentioned in Chapt. VII.B, one possible identification for the 25-cm^{-1} mode observed in lysozyme is that it is related to the hinge-bending motion. Normal-mode simulation results suggest that torsional vibrations in β-sheets occur in the range 15 to 40 cm^{-1} and that the α-helix torsional vibrations are in range 55 to 100 cm^{-1}.[136a] This corresponds to values found in measurements of α-helical and β-sheet polyalanine.[465,466,475]

An area where significant progress has been made recently is in studies of the incoherent inelastic neutron scattering of proteins.[29a,476] In this method low-energy neutrons with wavelengths on the order of 1 to 10 Å and energies on the order of kT ($T = 300$ K) interact inelastically with the protein. The scattered intensity is measured as a function of both energy and momentum transfer and can give simultaneous information about the temporal and spatial characteristics of the atomic motions. In hydrogen-rich materials, such as hydrocarbons and proteins, the scattering is primarily incoherent, due to the large incoherent cross section of the hydrogen atom. This means that the observed scattering is dominated by independent contributions arising from the self-correlation functions for the time-dependent hydrogen positions. Spectrometers now in use permit the study of diffusional, rotational, and internal motions in the range 10^{-7} to 10^{-13} s.

An early measurement of inelastic neutron scattering was made on polyalanine in the α-helical and β-sheet configurations.[475] The principal peak observed in that spectrum (~ 230 cm^{-1}) was assigned to the torsional oscillation of the methyl sidechain of alanine, in correspondence with methyl group torsions in methyl alcohol[477] and other methyl-substituted systems.[478]

A number of protein spectra have now been determined and work in this area has been stimulated by the existence of a time of flight spectrometer (IN6) at the Institut Laue-Langevin that is capable of measuring improved protein spectra.[391,476,479,480] The resuting spectra for proteins in solution have a rather similar smooth shape on a neutron time-of-flight scale. There is a strongly broadened quasi-elastic peak and then continuously increasing inelastic scattering from about 10 cm^{-1} out to 400 cm^{-1}, where a relatively abrupt drop in scattering intensity occurs. There appear to be some more intense bands superposed on the background scattering, particularly in the region 150 to 250 cm^{-1}. Measurements for lysozyme[479,480] indicate that the spectrum changes in the observed range when substrate is bound. It has been suggested that this change is correlated with the hinge-bending motion.

To obtain further insight into the meaning of the inelastic neutron spectra, it is necessary to have specific theoretical models with which to compare the experimental results. In the harmonic approximation it is possible to calculate the incoherent inelastic neutron spectrum; i.e., the neutron scattering cross section for the absorption or emission of a specific number of phonons can be obtained with the exact formulation of Zemach and Glauber.[481] A full multiphonon inelastic spectrum can be evaluated by use of Fourier transform techniques.[482] The availability of the normal-mode analysis for the BPTI[136] has made possible detailed one-phonon calculations[483] for this system; the one-phonon spectrum arises from transitions between adjacent vibrational levels and is the dominant contribution to the scattering at low frequencies for typical experimental conditions.[483] The calculated one-phonon neutron energy gain spectrum, convoluted with resolution broadening, has a number of peaks in the range 7 to 50 cm^{-1} that arise from the non-uniformity in the density of vibrational states. These peaks are not present in the observed spectrum.[484] An improved calculation of the normal mode frequencies and the resulting neutron spectrum does not show these low frequency peaks.[485] In addition, the experimental spectrum shows more intensity in the range 100 to 300 cm^{-1} than do the calculated results. It is likely that multiphonon corrections as well as the inclusion of the methyl group motions (neglected in the extended atom treatment used for the normal-mode model since a methyl group is approximated as a sphere with no torsional degree of freedom) would improve the agreement between the calculated and measured results; a calculation of the spectrum with an all-hydrogen model (see Chapt. III) does show increased intensity in the higher frequency region.[485] A fuller understanding of the low-frequency motions in proteins is expected from such combined theoretical and experimental studies.

E.   ELECTRON SPIN RELAXATION

Electron spin relaxation in paramagnetic systems is dominated by the vibrational modes that modulate the crystal field at the ion where the electron spin is localized.[486] Measurements in proteins have been made and attempts to analyze the results in terms of the vibrational modes have been described.[487] Of particular interest are measurements of the electron spin relaxation rate $(1/T_1)$ of low-spin ferric iron in a number of heme and iron-sulfur proteins.[488,489] For the temperature range between 4 and 20 K, the relaxation rate is dominated by a two-phonon (Raman) process with a temperature dependence that deviates significantly from the T^9 power law expected[490] and found experimentally[491] in ordinary three-dimensional solids. For a Debye temperature[164] much higher than the temperature of interest, the contribution of the Raman process is given by

$$1/T_1 = \int_0^\infty \frac{\omega^4 [g(\omega)]^2 \exp(h\omega/kT)}{[\exp(h\omega/kT) - 1]^2} \, d\omega \qquad (114)$$

where $g(\omega)$ is the density of vibrational states with frequency ω. Equation 114 leads to a temperature dependence of the form $T^{5+2\gamma}$ for a density of states that can be expressed by the simple power law ω^γ. Thus, for an ordinary three-dimensional solid, $\gamma = D - 1 = 2$, where D is the physical dimension and $1/T_1 \propto T^9$, as observed. For low-spin ferric iron proteins, an apparent temperature dependence of $T^{5.6}$ to $T^{6.3}$ has been measured.[488,489] From Eq. 114, this result corresponds to $\gamma = 0.3$ to 0.65, or $D_{\text{eff}} = 1.3$ to 1.65, which has been interpreted as the "fractal" geometric dimension of the polypeptide chain. However, this interpretation has been shown to be based on a misinterpretation of the theory.[492]

Since the electron spin relaxation rate in Eq. 114 is dominated by the low-frequency modes, an understanding of the origin of the unusual temperature dependence for proteins is of particular interest as a probe of the motional properties. Use of the normal-mode calculation for BPTI[136] yields an exponent $\gamma = 0.35$ for the density of states in the frequency range of interest (0 to 50 cm^{-1}).[493] This is in accord with the experimental estimates, although BPTI is not one of the proteins studied experimentally. However, the inelastic neutron data,[29a] as well as normal-mode calculations,[136a] suggest that the frequency dependence of $g(\omega)$ is similar for different proteins in the low-frequency range.

To understand the origin of the "fractal" density of states, a simplified model for the low-frequency vibrations of proteins was developed.[493] It is based on the effective medium approximation[494] that has been used successfully to describe other disordered (aperiodic) solids. The essential element of the model is its focus on the nature of the interactions among amino acid residues. In this simplified picture each amino acid has the possibility of four "strong" interactions, two with its bonded neighbors along the chain and two with other amino acids involving its hydrogen bonding (C=O and N—H) groups. These four interactions, if saturated, would lead to a protein that behaves approximately as a two-dimensional object ($D_{eff} = 2$) in the low-frequency range. However, because many of the hydrogen-bonding interactions are short range (i.e., those in α-helices) and therefore do not effect the low-frequency modes, the actual value of D_{eff} for most proteins is expected to be less than 2 ($1 \le D_{eff} \le 2$). In fact, the values for $g(\omega)$ from an effective medium approximation calculation with force constants estimated from the full protein potential energy expression (Eq. 6) yield D_{eff} between 1.3 and 1.6 (γ between 0.3 and 0.6) for the proteins myoglobin, cytochrome c_{551} and ferredoxin, three molecules that have been studied experimentally. Low-temperature specific-heat measurements for solid polypeptide chains[495] are also in accord with the model. They depend on the density of states in the low-frequency region and can be interpreted in terms of a one-dimensional model ($D_{eff} = 1$) for α-helical poly-L-alanine and a two-dimensional model ($D_{eff} = 2$) for β-sheet poly-L-alanine.

F. HYDROGEN EXCHANGE

Exchange experiments for amide (NH) and other exchangeable protons (OH, SH) were historically the first results that focused on the internal mobility of proteins.[496] Much work has been done in this area over the intervening years,[31,40] and it has been shown that the exchange rates for NH protons in native proteins can be as much as eight orders of magnitude slower than in small model systems, where the peptide group is exposed to solvent. Detailed analysis of these results has been hampered by the lack of data concerning the exchange rates for specific protons. A variety of exchange mechanisms have been proposed[31,33] and controversy has resulted because definitive evidence concerning any one of them has been difficult to obtain.

In recent years nuclear magnetic resonance studies in solution and coherent neutron diffraction measurements on crystals have provided results for the exchange rates of specific protons in several proteins. For lysozyme[497] and the bovine pancreatic trypsin inhibitor,[498,499] the exchange rates of nearly all N—H protons have been determined. Neutron data for crystals have provided qualitative exchange data ("fast" or "slow" exchange) for specific hydrogens in trypsin,[348] lysozyme,[500] and ribonuclease.[501] For lysozyme, where a detailed comparison has been made of the NMR and neutron diffraction results, the two methods are in good agreement.[502]

From measurements as a function of pH and temperature it is now clear that there are two limiting exchange mechanisms. In one of these, associated with conditions in which the protein is relatively unstable (high temperature, low pH), the exchange proceeds through an essentially denatured state. Evidence for this is provided by the fact that the excess activation energy (the difference between the measured activation energy and that required for model peptides) is close to that required for denaturation, and by the fact that the rates for protons in different regions of the protein approach the same value as the temperature is raised or the pH is lowered. In the other limit (low temperature, physiological pH), the exchange is a much more local phenomenon, with the specific rate for individual protons determined by a variety of environmental factors (e.g., whether or not the proton is hydrogen bonded, whether or not it is accessible, whether or not it is part of secondary structural elements). It has been shown that the calculated local electric field due to the protein charge distribution plays an important role in determining the pH dependence of the exchange rates.[503-505] This correlation of exchange rates with the calculated electrostatic field suggests that the rate-determining step takes place from a structure close to the native one used in the calculations.[504,505]

In spite of the detailed data now available, a full dynamical interpretation of hydrogen exchange is still lacking. A simulation study of BPTI[505a] found a correlation between the average lengths and fluctuations of hydrogen bonds and the measured exchange rates for a limited set of NH protons involved in the hydrogen bonds. When the more recent experimental results for all NH protons[505b] are compared with the calculations, there is little or no correlation. This is not surprising since the extrapolation required to go from the dynamics simulation (ps) to hydrogen exchange rates (hours) is very great. Such long-time phenomena are governed by significant activation energies and require special simulation methods for their analysis (e.g., activated dynamics, see Chapt. IV.E). For the hydrogen exchange process, the problem is particularly difficult because the reaction path or paths are likely to be complex. A study of the reaction-path problem has been made for the relatively rigid peptide ferrichrome, which contains several slowly exchanging hydrogens.[506] For six N—H hydrogens, it was found that the exchange rate correlated qualitatively with the solvent exposure generated by following the low-frequency normal modes that are expected to be thermally excited. It is possible that this type of approach could be extended to a more quantitative treatment of hydrogen exchange in other peptides and in proteins.

G. MÖSSBAUER SPECTROSCOPY

Mössbauer spectroscopy, based on the resonant absorption of nuclear gamma rays[507] has been used to provide data concerning the time scales and amplitudes of atomic motions in proteins. Such measurements are usually limited to the motion of the Mössbauer nucleus itself, although Rayleigh scattering of Mössbauer radiation[508] can yield results concerning the average atomic fluctuations of the entire system. Most protein studies have used ^{57}Fe as the Mössbauer nucleus. Very detailed dynamical investigations have been made for myoglobin,[385,509,510] and some data are available for other proteins, including cytochrome c[511] and ferritin.[512] Several aspects of Mössbauer spectroscopy make it of particular interest for obtaining information on dynamics. First, unlike X-ray diffraction, there is no static disorder contribution to the Lamb-Mössbauer factor, so that only the motion of the nucleus affects the observed spectrum. Second, the observed spectral line shape has measurable contributions over a range of times between approximately 1 and 100 ns, the latter being the gamma-ray emission lifetime (natural line width) of the ^{57}Fe nucleus, although faster motions can be estimated from the area of the narrow Mössbauer line;[385,509] and third, it is experimentally feasible to do measurements over a wide temperature range (e.g., the myoglobin work covers the range from 4.2 to 300 K).

For myoglobin it has been observed that the Lamb-Mössbauer factor, $-\ln f$, which is directly related to the mean-square fluctuations of the iron if the motion is harmonic ($f = \exp(-k \langle \Delta x^2 \rangle)$) where k is the momentum of the γ-quantum, increases linearly with temperature between 0 and 170 K and rises much more rapidly above 200 K. Further, a broad line appears in the Mössbauer spectrum at about 180 K and both its width and its intensity increase strongly with temperature; the linewidth reflects relaxation processes on the time scale $1 \leq t < 100$ ns. This behavior is correlated with the melting of the aqueous solvent that surrounds the protein even in a crystal and support the conclusion from molecular dynamics and normal-mode studies that the larger-scale displacements arise from low-frequency collective motions that involve the protein surface (see Chapt. VI.A).

The X-ray temperature factor for Fe corresponds to larger values for the mean-square displacements than the Mössbauer results, a result that is in accord with the presence of longer time and static disorder contributions to the former. One estimate of the disorder contribution is obtained by extrapolating the temperature-dependent results to 0 K; the Mössbauer value of $\langle \Delta x^2 \rangle$ ($\langle \Delta x^2 \rangle = \frac{1}{3} \langle \Delta r^2 \rangle$) goes to zero and the X-ray value is approximately 0.03 to 0.05 Å^2. Detailed analysis of the Mössbauer line shapes[514,515] indicates that there are significant contributions to the iron atom displacement from processes on the time scale 1 to 100 ns at room temperature. Although there is considerable uncertainty in the result, it has been estimated[385] that 40% of the observed $\langle \Delta x^2 \rangle$ value at 300° K arises from motions on this time scale. The myoglobin simulation[516] yields $\langle \Delta x^2 \rangle = 0.057$ Å^2 for the iron on a picosecond time scale. Since the Mössbauer estimate is 0.065 Å^2, there must be slower motions on the nanosecond time scale that involve multiple wells (potential minima) separated by significant barriers. In fact, such descriptions of the iron motions have been given in phenomenological treatments of the Mössbauer experiments.[515] No attempt was made to specifically characterize the motions, although it appears from the simulations that the entire heme group is involved.[516]

Based on the Mössbauer experiments, it has been argued[385] that the time scale for *all* atomic fluctuations calculated in molecular dynamics simulations are in error by several orders of magnitude (i.e., that the time scale should be 1 to 100 ns instead of the calculated time scale of 50 to 100 ps). It is more likely that the iron motion has a significant ps contribution, as found in molecular dynamics simulations, but that there are additional contributions in the nanosecond range due to barrier crossing between multiple wells.[230] Any discrepancy between the calculated $\langle \Delta x^2 \rangle$ and that estimated from the measurements can be explained by the possibility that vacuum picosecond simulations yield fluctuations that are somewhat too large (a reduction of 35% would be required). Also, there are possible errors in the interpretation of experiments introduced by treating the motions as harmonic and isotropic.[516] As has been pointed out in the analysis of X-ray temperature factors (see this chapter, Sect. A above), motions governed by multiple potential wells are generally underestimated by refinement procedures. Thus, within the present uncertainties of both the theory and the experiments, there is no inconsistency between the X-ray and Mössbauer measurements, on the one hand, and the molecular dynamics simulations, on the other.

H. PHOTODISSOCIATION AND REBINDING KINETICS

An important tool for studying protein dynamics is provided by photoisomerization[470] and photodissociation experiments.[517] The protein that has been studied in greatest detail is myoglobin,[227,318] which is of particular interest from the viewpoint of protein dynamics because the exit of the photodissociated ligand and its reentry into the heme pocket is mediated by sidechain fluctuations (see Chapt. VI.B.2). The photodissociation experiment perturbs the system from its equilibrium state (liganded in this case) and permits one to follow the return to equilibrium. The essence of the experiment is to photodissociate the ligand (CO in the most detailed studies) by a saturating flash at $t = 0$ and to monitor the rebinding of the ligand. The fraction of myoglobin molecules $N(t)$ that remain unliganded as a function of time t is determined by electronic absorption spectroscopy in the Soret region. Data collected over a wide range of temperatures (4 to 320 K) have shown rebinding on time scales of 10^{-6} to 10^3 s. In addition to temperature, the effects of variation in the pressure, solution viscosity, and different ligands have also been examined.[30] Analysis of a large body of data has led to a model[518] in which the rebinding process is separated conceptually into three steps: entrance of the ligand from the solvent into the protein (this is prevented at temperatures below 200 K), penetration of the ligand through the protein into the heme pocket (which was discussed in Chapt. VI.B.2), and the chemical rebinding to the iron from the heme pocket. A striking result is that at low temperatures (40 to 180 K), where only geminate recombination occurs, the fraction $N(t)$ that remains unliganded does not decay exponentially with time but instead is best described by a power law [i.e., $N(t) \simeq (1 - t/t_0)^n$, where n and t_0 are parameters that depend on the temperature]. This nonexponential behavior has been associated with the chemical rebinding to the heme. An explanation of this result[227] suggests that at low temperatures, the individual protein molecules are trapped in slightly different conformations (substates) and that the barrier for the rebinding step is different in each of these conformations. The required distribution of barrier heights, $g(E)$, with a concomitant distribution of rebinding rate constants ($k = Ae^{-E/RT}$, where A is assumed constant for simplicity) yields a rebinding rate $N(t)$ $[N(t) = \int g(E)e^{-kt}dE]$, that has the form of a power law. By fitting the experimental data for different ligands and different proteins, the distribution function $g(E)$ has been evaluated for each case; the widths range over 5 kcal/mol, with each system (protein and ligand) having a different distribution.[30] At higher temperatures ($T > 210$ K), where exponential rebinding is observed, the transitions among the substates are presumed to be fast relative to the rebinding process, so that a single average activation energy governs the kinetics. Alternative explanations of the power-law dependence have been given (e.g., in terms of coupled Brownian oscillator models),[519,520] but it appears that they are not able to explain all of the data.[518] As discussed in Chapt. VII.A, a recent analysis[230] of a myoglobin molecular dynamics trajectory demonstrated that substates are present and suggests that they would be frozen in at low temperatures.

A variety of additional measurements have been made on the photolysis behavior of myoglobin and the related protein, hemoglobin. These include room-temperature transient electronic absorption studies in the nanosecond range,[521] nanosecond-to-picosecond resonance Raman spectroscopy,[522,523] and low-temperature transient electronic absorption measurements in regions other than the Soret band.[524]

The available data for myoglobin provide insights into the photodissociation and rebinding process that demonstrates its inherent complexity. Based on such information involving different temperatures and time scales, a hierarchical model for protein internal motions has been introduced.[524] The dissociation process is conceptualized as a "protein quake" in which structural changes propagate outward in earthquake-like "waves" from the iron after the iron-ligand bond is broken by the photodissociation flash. Further, the protein is presumed to have a hierarchy of microstates (substates) that have a "tree" structure (substates divided into subsubstates, and so on) with a succession of "branches" characterized by smaller differences in position and smaller magnitudes for the intervening energy barriers. Data from the experiments are used to provide evidence for the existence of at least four such tiers of substates.[524,525] However, it is not clear as yet whether the hierarchical model provides a correct description of the protein potential surface and the resulting protein motions.

CHAPTER XII

CONCLUDING DISCUSSION

In 1977 the first paper presenting a detailed molecular dynamics simulation of a small protein was published.[15] In the ten years since then, there has been an explosive growth in the research concerned with theoretical, as well as experimental, studies of the internal motions of proteins. Dynamic phenomena have been explored in depth for a variety of proteins and peptides (e.g., protein hormones and inhibitors, transport and storage proteins, enzymes). The magnitudes and time scales of the atomic motions have been delimited theoretically and related to a variety of experimental measurements, including NMR, X-ray diffraction, electronic absorption, fluorescence depolarization, Raman and infrared spectroscopy, and inelastic neutron scattering. Techniques have been introduced to extend theoretical methods from the subnanosecond time range accessible to standard molecular dynamics simulations to much longer time scales for certain processes by the use of stochastic, activated, harmonic, and simplified model dynamics. Further, the effect of solvent was introduced by stochastic dynamic techniques or accounted for explicitly in full dynamic simulations, in some cases for the crystal environment. Concomitantly, experimental approaches to the internal motions have been improved and developed and a wealth of data is being accumulated. The resulting interplay between theory and experiment is an essential element in the present vitality of the field of protein dynamics.

On the subnanosecond time scale our basic knowledge of protein motions is almost complete; that is, the types of motion that occur have been determined, their characteristics evaluated and the factors responsible for their properties delineated. Simulation methods have shown that the structural fluctuations in proteins are sizable; particularly large fluctuations are found where steric constraints due to molecular packing are small (e.g., in the exposed sidechains and external loops), but substantial mobility is also found in the protein interior. Local atomic displacements are correlated in a manner that tends to minimize disturbances of the global structure of the protein. This often leads to fluctuations larger than would be permitted in a rigid polypeptide matrix.

Motions on a longer time scale are being explored, but our present understanding is more limited. When the motion of interest can be described in terms of a reaction path (e.g., hinge-bending, local activated event), methods exist for determining the nature and rate of the process. Harmonic and simplified model dynamics, as well as reaction-path calculations, can provide information on slower processes. Motions that are slow owing to their complexity (e.g., they involve large-scale structural changes) are most difficult to treat theoretically. Stochastic dynamics methods have been particularly useful for such problems. An example is provided by simulations of simplified models of protein folding, a process that is very difficult to treat in full atomic detail. Also, faster computers permit the extension of simulation times to nanosecond periods and longer. This is the time range of many experimental methods in current use (e.g., fluorescence depolarization, nuclear magnetic resonance and Mössbauer spectroscopy) as well as of certain biological phenomena. Application of theoretical methods to analyze the motions probed by the experiments is providing a more complete understanding of the phenomena.

The influence of solvent on the structure, dynamics, and thermodynamics of proteins and polypeptides has been shown to be significant in many cases, although the subnanosecond internal fluctuations appear to be little altered. Larger-scale motions, which involve the protein surface and often occur on longer time scales, are sensitive to solvent damping. Conventional simulations of proteins in the presence of solvent (i.e., periodic boundary molecular dynamics) have shown such global effects. On a more local level, the solvent contribution to the potential of mean force alters the relative populations of different conformers of amino acid sidechains in proteins and polypeptides. Insights concerning the role of solvation in the binding of substrate molecules and the function of enzymes have emerged through the application of specialized dynamics methods and analytic integral equation techniques.

From the accumulated information on protein motions some more general concepts and questions are beginning to emerge. The thermodynamics of small systems makes clear that individual protein molecules undergo significant fluctuations in their extensive properties at room temperature.[526] Thus, for a protein like myoglobin at room temperature, the root-mean-square fluctuation in the energy is on the order of 30 kcal/mol relative to a total kinetic energy of about 1×10^4 kcal/mol, and that in the volume is about 50 Å^3 relative to a total volume of about 1×10^5 Å^3.[19] As interesting as such general results may be, they tell us nothing about the structural character of the fluctuations nor about their time scale. It is the more detailed aspects of the fluctuations that are a primary concern of dynamical simulation and of many experiments. One fundamental question focuses on the relation between the magnitude of a displacement and the energy and time scale involved. Most esthetically pleasing is the idea that the length, time, and energy scale of the fluctuations are simply related in protein motions, as they are in homogeneous elastic solids. This idea is embodied in hierarchical descriptions that have been proposed for the dynamics of proteins.[524] The multiple conformations or substates of a native protein have been likened to a tree structure with successively smaller displacements and energy barriers of decreasing magnitude separating the various conformations as one moves from the trunk out to the branches and twigs. The evidence for such a hierarchy is limited. An alternative possibility, far less pleasing, is that the specifics of each motion determine its character and that no such simple general scaling relation holds. Thus very localized motions, such as a ring flip, can be very slow because they have high barriers. Conversely, highly delocalized motions, including those involving the whole protein, can be fast if they occur in a single multidimensional potential well. Since each motion then becomes a problem in itself, it is essential to focus on those of particular interest for detailed study. In this case, a conceptual approach to protein dynamics would have to include the complexity that is embodied in chemistry rather than the simplicity that is often assumed in physics. The aperiodic, inhomogeneous, and highly variable nature of protein structures is suggestive in this regard. It is, in fact, likely that specific structural features that differentiate one protein from another are crucial to certain motions that play a role in biological function.

In the theory of protein dynamics there are two primary directions where active research and significant progress can be expected in the near future. One concerns the more detailed examination of the dynamics and the thermodynamics of processes of biological interest, and the other, an improvement in approaches to longer-time dynamics. Extensions of the methodologies described in this volume, as well as the availability of faster computers, are beginning to yield information concerning longer-time phenomena. As to the former, there are many biological problems to which dynamical methods can be and are being applied and for which a knowledge of the dynamics forms an essential part of a complete understanding. It is useful to summarize some of these in what follows; certain biological applications that have already been examined by dynamical methods and some that have not, are included.

For the transport protein hemoglobin, there is more evidence concerning the role of motions than for any other protein. The tertiary and quaternary structural changes that occur on ligand binding and their relation to the allosteric mechanism are well documented (Chapts. VI.B.2 and VII.C). An important role of the quaternary structural change is to transmit information over a longer distance than could take place by tertiary structural changes alone; the latter are generally damped out over distances of a few angstroms, unless amplified by the displacement of secondary structural elements or domains. The detailed dynamics of the allosteric mechanism have yet to be determined; in particular, the barriers along the reaction path from the deoxy to the oxy structure have to be fully analyzed, and simulations of the tertiary and quaternary transition are needed. For the related storage protein, myoglobin, fluctuations in the globin have been shown to be essential to the binding process; that is, the protein matrix in the X-ray structure is so tightly packed that there is no low energy path for the ligand to enter or leave the heme pocket. Only through structural fluctuations in certain bottleneck regions can the barriers be lowered sufficiently to obtain the observed rates of ligand binding and release. Energy minimization and molecular dynamics have been employed to investigate the displacements involved and the resulting barrier magnitudes. Specialized dynamic studies are beginning to be used to analyze the activation entropies and rates of ligand motion across the barriers.

In many proteins and peptides, the transport of substances is *through* the molecule rather than via translation of the molecule as in hemoglobin. The most obvious cases are membrane systems, in which fluctuations are likely to be of great importance in determining the kinetics of transport. For channels that open and close (e.g., gramicidin), as well for active transport involving enzymes (e.g., ATPases), fluctuations, in some cases highly correlated ones, are likely to be involved. At present, structural details and experimental studies of the motions are lacking. Some dynamical analyses have been made recently, particularly for model systems like the gramicidin channel,[527,528] and additional results can be expected in the near future.

In electron-transport systems, such as cytochrome c, cytochrome c peroxidase[529] and the photosynthetic center[530,530a] protein flexibility is likely to play a role in the electron transfer.[530b] Evidence favors a vibronic tunneling mechanism for transport through the protein and between proteins,[531] although other mechanisms are not fully excluded. In vibronic tunneling, processes that would be energetically forbidden for rigid proteins become allowed if the energies for conformational distortions are available. Experimental data[532,533] suggest that the important fluctuations are characterized by an average frequency in the range 50 to 250 cm^{-1}, on the order of those associated with collective modes of proteins.[136] Since the transfer rate is a sensitive (exponentially decreasing) function of donor-acceptor distance, it may be modulated by surface sidechain displacements or other fluctuations that allow for the approach of the donor and acceptor.

For proteins involved in binding, flexibility, and fluctuations enter into both the thermodynamics and the kinetics of the reactions. For the rate of binding of two macromolecules (protein-antigen and antibody, protein-inhibitor and enzyme), as well as for smaller multisite ligands, structural fluctuations involving domains, sidechains, hydrogen-bonding groups, and so on, can lead to lowering of the free energy barriers. This has been suggested by the dynamic studies on ribonuclease A and its complexes and may be of general importance in biological processes. Further, the mechanism and rate of the binding process are ideal subjects for dynamical analysis. Brownian dynamic simulations of ligand binding and its dependence on solution conditions are playing an important role in this area.

The relative flexibility of the free and bound ligand, as well as of the protein, must be considered in the overall thermodynamics of the binding reaction. If the free species has considerable flexibility and fluctuations are involved in the binding step, it is likely that the bound species will be less flexible and a significant entropic destabilization will result. Thus for strong binding in cases where the rate is not important, relatively rigid species are desirable. This would reduce the conformational entropy decrease and could lead to a very favorable free energy of binding if there is high complementarity in the two binding states. However, flexibility of the bound species can have a stabilizing effect since it partly compensates for the loss of translational and rotational entropy on binding. Conversely, the entropy loss of binding a flexible substrate or the rigidification of a protein on substrate binding can be used to reduce the binding constant even when strong, highly specific enthalpic interactions are present. This may be important in enzyme reactions where relatively weak binding can serve to obtain rapid substrate release. The required balance between flexibility and rigidity is determined by the function of the binding in each case.

Dynamical techniques can be employed to determine entropies and free energies. Recent applications of simulation methods in the framework of thermodynamic perturbation theory have demonstrated the possibility of evaluating free-energy differences encompassing a wide range of interesting phenomena. Examples include the difference in binding free energy of two closely related ligands, and the difference in stability of two closely related proteins. The calculations have also made clear the importance of accurate potential functions for the quantitative calculation of the relatively small differences in free energy that often play a role in such phenomena.

In the function of proteins as catalysts, there is the greatest possibility of contributions from internal motions. The role of structural changes induced by the binding of the substrate has been discussed. In addition to cooperative effects caused by quaternary alterations, a variety of results can arise from the perturbation of the tertiary structure. One example is the ordered binding of several substrates (or effectors and substrates), with the first molecule to bind altering the local conformation so as to increase or decrease the subsequent binding of other molecules. The occurrence of larger-scale changes, such as the closing of active-site clefts by substrate binding, as in certain kinases (Chapt. VII.B), has been interpreted in terms of catalytic specificity, alteration of the environment of the substrate, and exclusion of water that could compete with the enzymatic reaction. In large enzymes with more than one catalytic site or in coupled enzyme systems, conformational freedom may be important in moving the substrate along its path from one site to the next. Many of these processes, for which structural data are available, are ready for the application of dynamical methods.

The flexibility of the substrate-binding site in enzymes can result in effects corresponding to those already considered in receptor binding. For enzyme catalysis, the enhanced binding of a substrate with its geometry and electron distribution close to the transition state requires conformational fluctuations. Entropic effects are also likely to be of significance, both with respect to solvent release on substrate binding and possible changes in vibrational frequencies that alter the conformational entropy of the bound system in the enzyme-substrate complex or in the transition state. The inactivity of enzyme precursors, as in trypsinogen (Chapt. VIII.C), can result from the presence of excess conformational freedom in residues involved in forming the active site. The entropic cost of constraining them in the proper geometry for interacting with the substrate may be so high that the activity is significantly reduced relative to that of the normal enzyme where the same residues are held in place more rigidly. Simulation studies of such phenomena have shown how to elucidate the details of the structural transition involved. As to the time dependence of fluctuations and structural alterations, there are a variety of possibilities to be considered. In the binding of reactants and release of products, the time course of fluctuations in the enzyme could interact with the motion of the substrate; e.g., the opening and closing fluctuations of active-site clefts may be modified by interactions with the substrate as it enters or leaves the binding site. Also, the role of solvent in substrate "mimicry" (i.e., in directing and controlling specific active conformations) and in the modulation of the local environment may be important.

Fluctuations could play an essential role in determining the effective barriers for the enzyme catalyzed reactions. If the substrate is relatively tightly bound, the local fluctuation in the enzyme could couple to the substrate in such a way as to reduce the barriers significantly. If such coupling effects exist, specific structures could have developed through evolutionary pressure to introduce directionality and enhance the required fluctuations. Frictional effects that occur in the crossing of barriers in the interior of the protein could act to increase the transition-state lifetime and so alter the reaction rates relative to those predicted by conventional rate theory. Energy released locally in substrate binding may be utilized directly for catalyzing its reaction, perhaps by inducing certain fluctuations. Whether such an effect occurs would depend on the rate of dissipation of the (mainly) vibrational energy and the existence of patterns of atoms and interactions to channel the energy appropriately. It will be of great interest to determine whether any of the rather speculative possibilities outlined here for the role of the energy and directionality of structural fluctuations in enzymatic reactions can be documented theoretically or experimentally for specific systems.

A wide range of biological problems involving proteins, not to mention nucleic acids, polysaccharides, and membrane lipids, is ready for study, and exciting new results can be expected as dynamical methods are applied to them. In the coming years, we shall learn how to calculate meaningful rate constants for enzymatic reactions, ligand binding, and many of the other biologically important processes mentioned above. The role of flexibility and fluctuations will be understood in much greater detail. It should become possible to determine the effects on the dynamics and thermodynamics of changes in solvent conditions and protein amino acid sequence. As the predictive powers of the theoretical approaches increase, applications will be made to practical problems arising in areas such as drug design, genetic engineering and industrial enzyme technology.

REFERENCES

1. E. Schrödinger, *What Is Life?* Cambridge University Press, Cambridge, 1945.
2. C. Tanford, *The Hydrophobic Effect,* 2nd ed., Wiley, New York, 1980, p. 142.
3. D. C. Phillips, in *Biomolecular Stereodynamics,* II, R. H. Sarma, ed., Adenine Press, Guilderland, N.Y., 1981, p. 497.
4. J. A. Hirschfelder, H. Eyring, and B. Topley, *J. Chem. Phys.* **4,** 170 (1936).
5. M. Karplus, R. N. Porter, and R. D. Sharma, *J. Chem. Phys.* **43,** 3259 (1965).
6. R. N. Porter, *Ann. Rev. Phys. Chem.* **25,** 317 (1974).
7. R. B. Walker and J. C. Light, *Ann. Rev. Phys. Chem.* **31,** 401 (1980).
8. G. C. Schatz and A. Kuppermann, *J. Chem. Phys.* **62,** 2502 (1980).
9. R. L. Whetten, G. S. Ezra, and E. R. Grant, *Ann. Rev. Phys. Chem.* **36,** 277 (1985).
10. B. J. Alder and T. E. Wainwright, *J. Chem. Phys.* **31,** 459 (1959).
11. A. Rahman, *Phys. Rev.* **A136,** 405 (1964).
12. F. H. Stillinger and A. Rahman, *J. Chem. Phys.* **60,** 1545 (1974).
13. W. W. Wood and J. J. Erpenbeck, *Ann. Rev. Phys. Chem.* **27,** 319 (1976).
14. W. G. Hoover, *Ann. Rev. Phys. Chem.* **34,** 103 (1983).
15. J. A. McCammon, B. R. Gelin, and M. Karplus, *Nature (Lond.)* **267,** 585 (1977).
16. G. Careri, P. Fasella, and E. Gratton, *CRC Crit. Rev. Biochem.* **3,** 141 (1975).
17. G. Careri, P. Fasella, and E. Gratton, *Ann. Rev. Biophys. Bioeng.* **8,** 69 (1979).
18. A. Cooper, *Sci. Prog. Oxford* **66,** 473 (1980).
19. A. Cooper, *Prog. Biophys. Mol. Biol.* **44,** 181 (1984).
20. G. Weber, *Adv. Protein Chem.* **29,** 1 (1975).
21. M. Karplus, *Ber. Bunsen-ges. Phys. Chem.* **86,** 386 (1982).
22. M. Karplus and J. A. McCammon, *Ann. Rev. Biochem.* **53,** 263 (1983).
23. M. Levitt, *Ann. Rev. Biophys. Bioeng.* **11,** 251 (1982).
23a. R. M. Levy, *Ann. N.Y. Acad. Sci.* **482,** 24 (1987).
24. J. A. McCammon and M. Karplus, *Ann. Rev. Phys. Chem.* **31,** 29 (1980).
25. J. A. McCammon and M. Karplus, *Acc. Chem. Res.* **16,** 187 (1983).
26. J. A. McCammon, *Rep. Prog. Phys.* **47,** 1 (1984).
26a. B. M. Pettitt and M. Karplus, *Topics in Mol. Pharmacology* **3,** 75 (1986).
27. W. van Gunsteren and H. J. C. Berendsen, *Biochem. Soc. Trans.* **10,** 301 (1982).
28. G. R. Welch, B. Somogyi, and S. Damjanovich, *Prog. Biophys. Mol. Biol.* **39,** 109 (1978).
29. I. D. Campbell, C. M. Dobson, and R. J. P. Williams, *Adv. Chem. Phys.* **39,** 55 (1978).
29a. S. Cusack, *Comments Mol. Cell. Biophys.* **3,** 243 (1986).
30. P. G. Debrunner and H. Frauenfelder, *Ann. Rev. Phys. Chem.* **33,** 283 (1982).
30a. C. M. Dobson and M. Karplus, *Meth. in Enzymology* **131L,** 362 (1986).
31. S. W. Englander and N. R. Kallenbach, *Quart. Rev. Biophys.* **16,** 521 (1983).

32. F. R. N. Gurd and T. M. Rothgeb, *Adv. Protein Chem.* **33,** 73 (1979).

33. M. Karplus and J. A. McCammon, *CRC Crit. Rev. Biochem.* **9,** 293 (1981).

34. O. Jardetsky, *Acc. Chem. Res.* **14,** 291 (1981).

35. W. S. Bennett and R. Huber, *CRC Crit. Rev. Biochem.* **15**(4), 291 (1984).

36. W. L. Peticolas, *Methods Enzymol.* **61,** 425 (1979).

37. D. Ringe and G. A. Petsko, *Ann. Rev. Biophys. Bioeng.* **13,** 331 (1984).

37a. D. A. Torchia, *Ann. Rev. Biophys. Bioeng.* **13,** 125 (1984).

38. R. J. P. Williams, *Biol. Rev.* **54,** 389 (1979).

39. R. J. P. Williams, *Chem. Soc. Rev.* **9,** 325 (1980).

39a. G. Wagner and K. Wüthrich, *Meth. in Enzymology* **131L,** 307 (1986).

40. C. K. Woodward and B. D. Hilton, *Ann. Rev. Biophys. Bioeng.* **8,** 99 (1979).

41. M. O'Conner, R. Porter, J. Whelan, eds., *Mobility and Function in Proteins and Nucleic Acids* (Ciba Found. Symp., Vol. 93), Pitman, London, 1983.

42. J. Hermans, ed., *Molecular Dynamics and Protein Structure,* Polycrystal Book Service, Western Springs, Ill., 1985.

43. M. Karplus and J. A. McCammon, *Sci. Am.* **254**(4), 42 (1986).

44. J. A. McCammon and S. Harvey, *Dynamics of Proteins and Nucleic Acids,* Cambridge University Press, Cambridge, 1987.

45. R. E. Dickerson and I. Geis, *The Structure and Action of Proteins,* Harper & Row, New York, 1969.

46. G. Fermi and M. F. Perutz, *Haemoglobin and Myoglobin,* Oxford University Press, Oxford, 1982.

47. G. E. Schulz and R. H. Schirmer, *Principles of Protein Structure,* Springer, New York, 1979.

48. T. E. Creighton, *Proteins,* W. H. Freeman, New York, 1983.

49. J. C. Kendrew, R. E. Dickerson, B. E. Strandberg, K. G. Hart, D. R. Davies, D. C. Phillips, and V. C. Shore, *Nature (Lond.)* **185,** 422 (1960).

50. M. F. Perutz, *Nature (Lond.)* **167,** 1053 (1951).

51. (a) C. C. F. Blake, D. F. Koenig, G. A. Mair, A. C. T. North, D. C. Phillips, and V. R. Sarma, *Nature (Lond.)* **206,** 757 (1965); (b) C. C. F. Blake, L. N. Johnson, G. A. Mair, A. C. T. North, D. C. Phillips, and V. R. Sarma, *Proc. Roy. Soc. London* **B167,** 378 (1967).

52. F. C. Bernstein, T. F. Koetzle, G. J. B. Williams, E. F. Meyer, M. D. Brice, J. R. Rodgers, O. Kennard, T. Shimanouchi, and M. Tasumi, *J. Mol. Biol.* **112,** 535 (1977).

53. J. S. Richardson, *Adv. Protein Chem.* **34,** 167 (1981).

54. F. Sanger and E. O. P. Thompson, *Biochem. J.* **53,** 353 (1953).

55. C. J. Epstein, R. E. Goldberger, and C. B. Anfinsen, *Cold Spring Harbor Symp. Quant. Biol.* **28,** 439 (1963).

56. L. Pauling and R. B. Corey, *Proc. Roy. Soc. London* **B141,** 10 (1953).

57. J. A. Pople, B. T. Luke, M. J. Frisch, and J. S. Binkley, *J. Phys. Chem.* **89,** 2198 (1985).

58. G. N. Ramachandran, C. Ramakrishnan, and V. Sasisekharan, *J. Mol. Biol.* **7,** 95 (1963).

59. L. R. Pratt, C. S. Hsu, and D. Chandler, *J. Chem. Phys.* **68,** 4202 (1978).

60. H. A. Scheraga, *Adv. Phys. Org. Chem.* **6,** 103 (1968).

61. U. Burkert and N. L. Allinger, *Molecular Mechanics,* American Chemical Society, Washington, D.C., 1982.

62. K. D. Gibson and H. A. Scheraga, *Proc. Natl. Acad. Sci. USA* **58**, 420 (1967); L. Dunfield, A. Burgess, and H. Scheraga, *J. Phys. Chem.* **82**, 2609 (1978).

63. A. Hagler, E. Huber, and S. Lifson, *J. Am. Chem. Soc.* **96**, 5319 (1974).

64. M. Levitt, *J. Mol. Biol.* **168**, 595 (1983).

65. B. R. Brooks, R. Bruccoleri, B. Olafson, D. States, S. Swaninathan, and M. Karplus, *J. Comp. Chem.* **4**, 187 (1983).

66. S. J. Weiner, P. Kollman, D. Case, U. Singh, C. Ghio, G. Alagona, S. Profeta, and P. Weiner, *J. Am. Chem. Soc.* **106**, 765 (1984).

67. S. J. Weiner, P. Kollman, D. T. Nguyen, and D. Case, *J. Comp. Chem.* **7**, 230 (1986).

68. L. Nilsson and M. Karplus, *J. Comp. Chem.* **7**, 591 (1986).

69. M. Jackson, Ph.D. thesis, Harvard University, 1986.

70. L. D. Landau and E. M. Lifshitz, *Quantum Mechanics,* 3rd ed., Addison-Wesley, Reading, Mass., 1977, Sect. 89.

71. J. E. Lennard-Jones, *Proc. Roy. Soc. London* **A106**, 463 (1924).

72. E. S. Rittner, *J. Chem. Phys.* **19**, 1030 (1951).

73. R. S. Mulliken, *J. Chem. Phys.* **23**, 1833 (1955).

74. P. O. Lowdin, *Phys. Rev.* **97**, 1475 (1955).

75. W. Jorgensen, *J. Am. Chem. Soc.* **103**, 335 (1981); **103**, 341 (1981).

76. (a) S. R. Cox and D. E. Williams, *J. Comp. Chem.* **2**, 304 (1981); (b) U. C. Singh and P. Kollman, *J. Comp. Chem.* **5**, 129 (1984).

77. See, e.g., (a) D. L. Beveridge and M. Mezei in Ref. 42, p. 166; (b) K. Mehrotra and D. L. Beveridge, *J. Am. Chem. Soc.* **107**, 2239 (1985).

78. B. R. Gelin, Ph.D. thesis, Harvard University, 1976.

79. C. Tanford and J. G. Kirkwood, *J. Am. Chem. Soc.* **79**, 5333 (1957).

80. B. M. Pettitt, *J. Chem. Phys.* (submitted).

81. P. M. Morse, *Phys. Rev.* **34**, 57 (1929).

82. L. Pauling, *Nature of the Chemical Bond,* Cornell University Press, Ithaca, N.Y., 1939.

83. H. C. Urey and C. A. Bradley, *Phys. Rev.* **38**, 1969 (1931).

84. B. M. Pettitt and M. Karplus, *J. Am. Chem. Soc.* **107**, 1166 (1985).

85. P. J. Rossky, M. Karplus, and A. Rahman, *Biopolymers* **18**, 825 (1979).

86. A. T. Hagler, D. J. Osguthorpe, P. Dauber-Osguthorpe, and J. C. Hempel, *Science* **227**, 1309 (1985).

87. J. O. Hirschfelder, L. Curtiss, and R. B. Bird, *Molecular Theory of Gases and Liquids,* Wiley, New York, 1954.

88. E. B. Wilson, J. C. Decius, and P. C. Cross, *Molecular Vibrations,* Dover, New York, 1980.

89. W. C. Swope and H. C. Andersen, *J. Phys. Chem.* **88**, 6548 (1984).

90. L. Verlet, *Phys. Rev.* **159**, 98 (1967).

91. C. W. Gear, *Numerical Initial Value Problems in Ordinary Differential Equations,* Prentice-Hall, Englewood Cliffs, N.J., 1971.

92. W. F. van Gunsteren and H. J. C. Berendsen, *J. Mol. Biol.* **176**, 559 (1984).

93. D. N. Theodorou and U. W. Suter, *J. Chem. Phys.* **82**, 955 (1985).

94. H. C. Andersen, *J. Chem. Phys.* **72**, 2384 (1980).

95. J. P. Rykaert and G. Ciccotti, *J. Chem. Phys.* **78**, 7368 (1983).

96. H. J. C. Berendsen, J. P. M. Postma, W. F. van Gunsteren, A. DiNola, and J. R. Haak, *J. Chem. Phys.* **81**, 3684 (1984).

97. D. W. Wood, in *Water: A Comprehensive Treatise,* Vol. 6, F. Franks, ed., Plenum Press, New York, 1979, p. 279.

98. G. Ciccotti and A. Tenenbaum, *J. Stat. Phys.* **23**, 767 (1980).

99. C. L. Brooks III, A. Brünger, and M. Karplus, *Biopolymers* **24**, 843 (1985).

100. M. Berkowitz and J. A. McCammon, *Chem. Phys. Lett.* **90**, 215 (1982).

101. C. L. Brooks III and M. Karplus, *J. Chem. Phys.* **79**, 6312 (1983).

102. A. Brünger, C. L. Brooks III, and M. Karplus, *Chem. Phys. Lett.* **105**, 495 (1984).

103. A. Brünger, C. L. Brooks III, and M. Karplus, *Proc. Natl. Acad. Sci. USA* **82**, 8458 (1985).

104. S. A. Adelman and C. L. Brooks III, *J. Phys. Chem.* **86**, 1511 (1982).

105. J. C. Tully, *J. Chem. Phys.* **73**, 1975 (1980).

106. S. A. Adelman, *Adv. Chem. Phys.* **53**, 61 (1983).

107. G. Ciccotti and J.-P. Ryckaert, *Mol. Phys.* **40**, 141 (1980).

108. C. L. Brooks III and M. Karplus, *J. Mol. Biol.* (in press).

109. (a) D. Chandler, *J. Phys. Chem.* **88**, 3400 (1984); (b) L. R. Pratt and D. Chandler, *J. Chem. Phys.* **67**, 3683 (1977); **73**, 3430 (1980); **73**, 3434 (1980).

110. C. L. Brooks III and M. Karplus (to be published).

111. A. Belch and M. Berkowitz, *Chem. Phys. Lett.* **113**, 278 (1985).

112. F. Hirata and P. J. Rossky, *Chem. Phys. Lett.* **83**, 329 (1981).

113. F. Hirata, P. J. Rossky, and B. M. Pettitt, *J. Chem. Phys.* **78**, 4133 (1983).

114. D. Zichi and P. J. Rossky, *J. Chem. Phys.* **84**, 1712 (1986).

115. B. M. Pettitt and M. Karplus, *Chem. Phys. Lett.* **121**, 194 (1985).

115a. B. M. Pettitt and M. Karplus, *Chem. Phys. Lett.* **136**, 383 (1987).

116. J. T. Hynes, in *The Theory of Chemical Reaction Dynamics,* Vol. 4, M. Baer, ed., CRC Press, Boca Raton, Fla., 1985, p. 171.

117. D. Case and M. Karplus, *J. Mol. Biol.* **132**, 343 (1979).

118. G. H. Snyder, R. Rowan, S. Karplus, and B. D. Sykes, *Biochemistry* **14**, 3765 (1975).

119. G. Wagner, A. DeMarco, and K. Wüthrich, *Biophys. Struct. Mech.* **2**, 139 (1976).

120. R. Hetzel, K. Wüthrich, J. Deisenhofer, and R. Huber, *Biophys. Struct. Mech.* **2**, 159 (1976).

121. J. R. Knowles and W. J. Albery, *Acc. Chem. Res.* **10**, 105 (1977).

122. J. A. McCammon and M. Karplus, *Proc. Natl. Acad. Sci. USA* **76**, 3585 (1979).

123. J. A. McCammon and M. Karplus, *Biopolymers* **19**, 1375 (1980).

124. S. H. Northrup, M. R. Pear, C. Y. Lee, J. A. McCammon, and M. Karplus, *Proc. Natl. Acad. Sci. USA* **79**, 4035 (1982).

125. J. Keck, *Discuss. Faraday Soc.* **33**, 173 (1962).

126. J. B. Anderson, *J. Chem. Phys.* **58**, 4684 (1973).

127. P. Pechukas, in *Dynamics of Molecular Collisions, Part B,* W. H. Miller, ed., Plenum Press, New York, 1976, p. 269.

128. C. H. Bennett, in *Diffusion in Solids,* J. J. Burton and A. S. Nowick, eds., Academic Press, San Francisco, 1975.

128a. R. Elber and M. Karplus, *Chem. Phys. Lett.* **138**, 375 (1987).

129. J. P. Valleau and G. M. Torrie, in *Statistical Mechanics Part A: Equilibrium Techniques*, B. J. Berne, ed., Plenum Press, New York, 1977, pp. 137, 169.

130. B. R. Gelin and M. Karplus, *Proc. Natl. Acad. Sci. USA* **72**, 2002 (1975).

131. D. Chandler, *J. Chem. Phys.* **68**, 2959 (1978).

131a. J. A. Montgomery, D. Chandler, and B. J. Berne, *J. Chem. Phys.* **70**, 4056 (1979).

132. (a) W. H. Miller, N. C. Handy, and J. E. Adams, *J. Chem. Phys.* **72**, 99; (b) R. A. Marcus, *J. Chem. Phys.* **49**, 2617 (1968).

132a. (a) J. E. Straub and B. J. Berne, *J. Chem. Phys.* **83**, 1138 (1985); (b) J. E. Straub, D. A. Hsu, and B. J. Berne, *J. Phys. Chem.* **89**, 5188 (1985).

132b. P. Bash, M. Field, and M. Karplus, *J. Am. Chem. Soc.* **119**, 8092 (1987).

133. N. Gō and H. Scheraga, *Macromolecules* **9**, 535 (1976).

134. R. M. Levy and M. Karplus, *Biopolymers* **18**, 2465 (1979).

135. N. Gō, T. Noguti, and T. Nishikawa, *Proc. Natl. Acad. Sci. USA* **80**, 3696 (1985).

136. B. R. Brooks and M. Karplus, *Proc. Natl. Acad. Sci. USA* **80**, 6571 (1983).

136a. M. Levitt, C. Sauder, and P. S. Stern, *J. Mol. Biol.* **181**, 423 (1985).

137. R. M. Levy, D. Perahia, and M. Karplus, *Proc. Natl. Acad. Sci. USA* **79**, 1346 (1982).

138. K. Irikura, B. Tidor, B. R. Brooks, and M. Karplus, *Science* **229**, 571 (1985).

139. M. Born and K. Huang, *Dynamical Theory of Crystal Lattices*, Clarenden, Oxford Press, 1954.

140. M. Karplus and J. N. Kushick, *Macromolecules* **14**, 325 (1981).

141. R. Levy, M. Karplus, J. N. Kushick, and D. Perahia, *Macromolecules* **17**, 1370 (1984).

142. O. Edholm and H. J. C. Berendsen, *Mol. Phys.* **51**, 1011 (1984).

143. D. L. Freeman and J. D. Doll, *J. Chem. Phys.* **82**, 462 (1985).

144. R. W. Hockney and J. W. Eastwood, *Computer Simulation Using Particles*, McGraw-Hill, New York, 1981.

145. D. Beeman, *J. Comp. Phys.* **20**, 130 (1976).

146. W. F. van Gunsteren and H. J. C. Berendsen, *Mol. Phys.* **34**, 1311 (1977).

147. W. F. van Gunsteren and H. J. C. Berendsen, *Mol. Phys.* **45**, 637 (1982).

148. W. F. van Gunsteren and M. Karplus, *Macromolecules* **15**, 1528 (1982).

149. D. L. Ermak and J. A. McCammon, *J. Chem. Phys.* **69**, 1352 (1978).

150. M. R. Pear, S. H. Northrup, J. A. McCammon, M. Karplus, and R. M. Levy, *Biopolymers* **20**, 629 (1981).

151. S. Y. Lee, M. Karplus, D. Bashford, and D. Weaver, *Biopolymers* **26**, 481 (1987).

152. J. A. McCammon, B. R. Gelin, M. Karplus, and P. G. Wolynes, *Nature (Lond.)* **262**, 325 (1976).

153. J. A. McCammon, P. G. Wolynes, and M. Karplus, *Biochemistry* **18**, 927 (1979).

154. A. Szabo, D. Shoup, S. H. Northrup, and J. A. McCammon, *J. Chem. Phys.* **77**, 4484 (1982).

155. S. A. Allison and J. A. McCammon, *J. Phys. Chem.* **89**, 1072 (1985).

156. S. L. S. Jacoby, K. S. Kowalik, and J. T. Pizzo, *Iterative Methods for Nonlinear Optimization Problems*, Prentice-Hall, Englewood Cliffs, N.J., 1972.

157. K. B. Wiberg, *J. Am. Chem. Soc.* **87**, 1070 (1965).

158. J. E. Williams, P. J. Stang, and P. V. R. Schleyer, *Ann. Rev. Phys. Chem.* **19**, 531 (1968).

159. M. J. D. Powell, *Math. Programming* **12**, 241 (1977).

160. R. H. Boyd, *J. Chem. Phys.* **49**, 2574 (1968).

160a. J. W. Ponder and F. M. Richards, *J. Mol. Biol.* **193**, 775 (1987).

160b. M. R. Pincus and H. A. Scheraga, *Biochemistry* **20**, 3960 (1981).

160c. P. Kollman, G. Wipff, and U. C. Singh, *J. Am. Chem. Soc.* **107**, 2212 (1985).

161. D. Hall and N. Pavitt, *J. Comp. Chem.* **5**, 441 (1984).

161a. M. Whitlow and M. M. Teeter, *J. Am. Chem. Soc.* **108**, 7163 (1986).

162. M. Karplus, *Ann. N.Y. Acad. Sci.* **439**, 107 (1985).

163. P. J. Rossky and M. Karplus, *J. Am. Chem. Soc.* **101**, 1913 (1979).

164. D. McQuarrie, *Statistical Mechanics,* Harper & Row, New York, 1975.

165. M. I. Page and W. P. Jencks, *Proc. Natl. Acad. Sci. USA* **68**, 1678 (1971).

166. I. Z. Steinberg and H. A. Scheraga, *J. Biol. Chem.* **238**, 172 (1963).

167. B. Tidor and M. Karplus (to be published).

168. H. L. Friedman and W. D. T. Dale, in *Statistical Mechanics, Part A: Equilibrium Techniques,* B. J. Berne, ed., Plenum Press, New York, 1977, p. 85.

169. F. Garisto, P. G. Kusalik, and G. N. Patey, *J. Chem. Phys.* **79**, 6294 (1983).

169a. H. Yu and M. Karplus, *J. Chem. Phys.* (in press).

170. J. P. Hansen and I. R. McDonald, *Theory of Simple Liquids,* Academic Press, New York, 1976.

170a. M. Mezei, P. K. Mehrotra, and D. L. Beveridge, *J. Am. Chem. Soc.* **107**, 2239 (1985).

170b. R. Zwanzig, *J. Chem. Phys.* **22**, 1420 (1954).

170c. C. H. Bennett, *J. Comput. Phys.* **19**, 267 (1975).

170d. B. J. Berne, in *Multiple Time Scales,* J. V. Brackloill and B. I. Cohen, eds., Academic Press, New York, 1985.

171. S. H. Northrup and J. A. McCammon, *Biopolymers* **19**, 1001 (1980).

172. R. Friesner and R. M. Levy, *J. Chem. Phys.* **80**, 4488 (1984).

173. M. Mezei, *Mol. Phys.* **47**, 1307 (1982).

174. J. P. M. Postma, H. J. C. Berendsen, and J. R. Haak, *Faraday Symp. Chem. Soc.* **17**, 55 (1982).

175. J. G. Kirkwood, *J. Chem. Phys.* **3**, 300 (1935).

176. J. D. Weeks, D. Chandler, and H. C. Anderson, *J. Chem. Phys.* **54**, 5237 (1971).

177. H. Callen, *Thermodynamics and an Introduction to Thermostatistics,* 2nd ed., Wiley, New York, 1985, p. 433.

178. D. Chandler, in *The Liquid State of Matter: Fluids, Simple and Complex,* E. W. Montroll and J. L. Lebowitz, eds., North-Holland, New York, 1982, p. 275.

179. R. A. Chiles and P. J. Rossky, *J. Am. Chem. Soc.* **106**, 6867 (1984).

180. C. L. Brooks, B. M. Pettitt, and M. Karplus, *J. Chem. Phys.* **83**, 5897 (1985).

181. C. G. Gray and K. E. Gubbins, *Theory of Molecular Fluids,* Oxford University Press, New York, 1984.

182. H. L. Friedman, *Ann. Rev. Phys. Chem.* **32**, 179 (1981).

183. B. M. Pettitt and P. J. Rossky, *J. Chem. Phys.* **77**, 1451 (1982).

184. G. Stell, G. N. Patey, and J. S. Hoeye, *Adv. Chem. Phys.* **48**, 183 (1981).

185. S. Singer and D. Chandler, *Mol. Phys.* **55**, 621 (1985).

186. D. Chandler, Y. Singh, and D. Richardson, *J. Chem. Phys.* **81**, 1975 (1984).

187. B. M. Pettitt and M. Karplus, in *Molecular Graphics and Drug Design*, A. S. V. Burgen, G. C. K. Roberts & M. S. Tute, eds., Elsevier Science Publishers, B.V., 1986.

188. S. J. Singer, R. A. Kuharski, and D. Chandler, *J. Phys. Chem.* **90**, 6015 (1986).

189. C. L. Brooks III, *J. Chem. Phys.* **86**, 5156 (1987).

190. R. M. Levy, R. P. Sheridan, J. W. Keepers, G. S. Dubey, S. Swaminathan, and M. Karplus, *Biophys. J.* **48**, 509 (1985).

191. T. Ichiye, B. Olafson, S. Swaminathan, and M. Karplus, *Biopolymers* **25**, 1909 (1986).

192. C. B. Post, B. R. Brooks, C. M. Dobson, P. Artymiuk, J. Cheetham, D. C. Phillips, and M. Karplus, *J. Mol. Biol.* **190**, 455 (1986).

193. W. van Gunsteren and M. Karplus, *Biochemistry* **21**, 2259 (1982).

194. S. H. Northrup, M. R. Pear, J. D. Morgan, J. A. McCammon, and M. Karplus, *J. Mol. Biol.* **153**, 1087 (1981).

195. B. Mao, M. R. Pear, J. A. McCammon, and S. H. Northrup, *Biopolymers* **21**, 1979 (1982).

196. T. Ichiye and M. Karplus, *Proteins* **2**, 236 (1987).

197. J. Kuriyan, G. Petsko, R. M. Levy, and M. Karplus, *J. Mol. Biol.* **190**, 227 (1986).

198. J. D. Morgan, J. A. McCammon, and S. H. Northrup, *Biopolymers* **22**, 1579 (1983).

199. S. Swaminathan, T. Ichiye, W. F. van Gunsteren, and M. Karplus, *Biochemistry* **21**, 5230 (1982).

200. M. Gō and N. Gō, *Biopolymers* **15**, 1119 (1976).

201. Y. Suezaki and N. Gō, *Biopolymers* **15**, 2137 (1976).

202. H. Hartmann, F. Parak, W. Steigemann, G. A. Petsko, D. Ringe Ponzi, and H. Frauenfelder, *Proc. Natl. Acad. Sci. USA* **79**, 4967 (1982).

203. R. M. Levy, A. R. Srinivasan, W. Olson, and J. A. McCammon, *Biopolymers* **23**, 1099 (1984).

204. B. R. Brooks and M. Karplus (to be published).

205. P. J. Flory, *Statistical Mechanics of Chain Molecules*, Wiley, New York, 1969.

206. J. A. McCammon, S. H. Northrup, M. Karplus, and R. M. Levy, *Biopolymers* **19**, 2033 (1980).

207. M. Karplus and J. A. McCammon, *Nature (Lond.)* **277**, 578 (1979).

208. J. M. Sturtevant, *Proc. Natl. Acad. Sci. USA* **74**, 2236 (1977).

209. B. R. Gelin and M. Karplus, *Biochemistry* **18**, 1256 (1979).

210. R. Zwanzig, *Ann. Rev. Phys. Chem.* **16**, 67 (1965).

211. B. J. Berne and G. D. Harp, *Adv. Chem. Phys.* **17**, 63 (1970).

212. S. Chandrasekhar, *Rev. Mod. Phys.* **15**, 1 (1943).

213. D. Enskog, in *Kinetic Theory*, Vol. 3, S. G. Brush, ed., Pergamon Press, Oxford, 1972, p. 125.

214. R. Gordon, *J. Chem. Phys.* **44**, 1830 (1966).

215. D. Chandler, *J. Chem. Phys.* **60**, 3500 (1974).

216. D. F. Calef and J. M. Deutch, *Ann. Rev. Phys. Chem.* **34**, 493 (1983).

217. A. Warshel and M. Karplus, *J. Am. Chem. Soc.* **94**, 5612 (1972).

218. B. R. Gelin and M. Karplus, *J. Am. Chem. Soc.* **97**, 6996 (1975).

219. H. A. Kramers, *Physica* **7**, 284 (1940).

220. J. A. McCammon, C. Y. Lee, and S. H. Northrup, *J. Am. Chem. Soc.* **105**, 2232 (1983).

221. R. O. Rosenberg, B. J. Berne, and D. Chandler, *Chem. Phys. Lett.* **75**, 162 (1980).

222. D. Bunker, *Theory of Elementary Gas Reaction Rates,* Pergamon Press, Oxford, 1966.

223. J. L. Skinner and P. G. Wolynes, *J. Chem. Phys.* **69**, 2143 (1978).

224. R. Pastor and M. Karplus (to be published).

225. T. Takano, *J. Mol. Biol.* **110**, 569 (1977).

226. S. E. V. Phillips, *Nature* **273**, 247 (1978).

226a. R. Elber and M. Karplus (to be published).

227. R. H. Austin, K. W. Beeson, L. Eisenstein, H. Frauenfelder, and I. C. Gunsalus, *Biochemistry* **14**, 5355 (1975).

228. D. A. Case and J. A. McCammon, *Ann. N.Y. Acad. Sci.* **482**, 222 (1987).

229. J. Äqvist, W. F. van Gunsteren, M. Leijonmarck, and O. Tapia, *J. Mol. Biol.* **183**, 461 (1985).

230. R. Elber and M. Karplus, *Science* **235**, 318 (1987).

231. C. Chothia and A. M. Lesk, *Trends Biochem.* **10**, 116 (1985).

232. C. Chothia, A. M. Lesk, G. G. Dodson, and D. C. Hodgkin, *Nature (Lond.)* **302**, 500 (1983).

233. A. M. Lesk and C. Chothia, *J. Mol. Biol.* **136**, 225 (1980).

234. C. M. Anderson, F. H. Zucher, and T. A. Steitz, *Science* **204**, 375 (1979).

235. L. N. Johnson, in *Inclusion Compounds,* J. L. Atwood, T. E. D. Davies, and D. D. MacNicol, eds., Academic Press, New York, 1983.

236. J. Janin and S. J. Wodak, *Prog. Biophys. Mol. Biol.* **42**, 21 (1983).

237. T. Imoto, L. N. Johnson, A. C. T. North, D. C. Phillips, and J. A. Rupley, *The Enzymes* **7**, 3rd. ed. 665 (1972).

238. P. G. Wolynes and J. A. McCammon, *Macromolecules* **10**, 86 (1977).

239. R. Bruccoleri, M. Karplus, and J. A. McCammon, *Biopolymers* **25**, 1767 (1986).

239a. B. R. Brooks and M. Karplus, *Proc. Nat. Acad. Sci. USA,* **82**, 4995 (1985).

240. C. Lanczos, *J. Res. Natl. Bur. Stand.* **45**, 255 (1950).

241. J. A. McCammon and M. Karplus, *Nature (Lond.)* **268**, 765 (1977).

242. V. T. Oi, T. M. Vuong, R. Hardy, J. Reidler, J. Dangl, L. A. Herzenberg, and L. Stryer, *Nature (Lond.)* **307**, 136 (1984).

243. B. Mao, M. R. Pear, J. A. McCammon, and F. A. Quiocho, *J. Biol. Chem.* **257**, 1131 (1982).

244. F. Colona-Cesari, D. Perahia, M. Karplus, H. Ecklund, C.-I. Branden, and O. Tapia, *J. Biol. Chem.* **261**, 15273 (1986).

245. M. E. Newcomer, B. A. Lewis, and F. A. Quiocho, *J. Biol. Chem.* **256**, 13218 (1981).

246. E. Eklund, B. Nordström, E. Zeppezauer, G. Söderlund, I. Ohlsson, T. Boiwa, B.-O. Söderberg, O. Tapia, and C.-I. Brändén, *J. Mol. Biol.* **102**, 27 (1976).

247. H. Eklund, J. P. Samama, L. Wallén, C.-I. Brändén, A. Akeson, and T. A. Jones, *J. Mol. Biol.* **146**, 561 (1981).

248. J. A. McCammon and S. H. Northrup, *Nature (Lond.)* **293**, 316 (1981).

249. S. H. Northrup, F. Zarrin, and J. A. McCammon, *J. Phys. Chem.* **86**, 2314 (1982).

250. S. Y. Lee and M. Karplus, *J. Chem. Phys.* **86**, 1883 (1987); *ibid* **86**, 1904 (1987).

251. S. C. Harrison, *Trends Biochem. Sci.* **3**, 3 (1978).

252. S. C. Harrison, *Biophys. J.* **32**, 139 (1980).

253. R. C. McDonald, T. A. Steitz, and D. M. Engelman, *Biochemistry* **18**, 338 (1979).

254. C. A. Pickover, D. B. McKay, D. M. Engelman, and T. A. Steitz, *J. Biol. Chem.* **254**, 11323 (1979).

255. D. C. Hanson, J. Yguerabide, and V. N. Schumaker, *Biochemistry* **20**, 6842 (1981).

255a. R. A. Mendelson, M. F. Morales, and J. Botts, *Biochemistry* **12**, 2250 (1973).

255b. J. Smith, S. Cusack, P. Poole, and J. L. Finney, *J. Biomol. Struct. Dynam.* **4**, 583 (1987).

256. T. Y. Morozova and V. N. Morozov, *J. Mol. Biol.* **157**, 173 (1982).

257. J. A. McCammon and P. G. Wolynes, *J. Chem. Phys.* **66**, 1452 (1977).

258. L. Genzel, R. Keilmann, T. P. Martin, G. Winterling, Y. Yacoby, H. Fröhlich, and M. Makinen, *Biopolymers* **15**, 219 (1976).

259. K. G. Brown, S. C. Erfurth, E. W. Small, and W. L. Peticolas, *Proc. Natl. Acad. Sci. USA* **69**, 1467 (1972).

260. D. W. Darnall and I. M. Klotz, *Arch. Biochem. Biophys.* **166**, 651 (1975).

261. E. R. Kantrowitz, S. C. Pastra-Landis, and W. N. Lipscomb, *Trends Biochem. Sci.* **5**, 124 (1980); **5**, 150 (1980).

262. Ref. 48, Chap. 9.

263. R. E. Dickerson and I. Geis, *Hemoglobin: Structure Function, Evolution and Pathology*, Benjamin/Cummings, Menlo Park, Calif., 1983.

264. H. F. Bunn and B. G. Forget, *Hemoglobin: Molecular, Genetic and Clinical Aspects*, W. B. Saunders, Philadelphia, 1986.

265. J. Baldwin and C. Chothia, *J. Mol. Biol.* **129**, 175 (1979).

266. F. A. Ferrone, A. J. Martino, and S. Basak, *Biophys. J.* **48**, 269 (1985).

267. B. R. Gelin, A. W.-M. Lee, and M. Karplus, *J. Mol. Biol.* **171**, 489 (1983).

268. E. R. Henry, J. H. Sommer, J. Hofrichter, and W. A. Eaton, *J. Mol. Biol.* **166**, 443 (1983).

269. A. Szabo and M. Karplus, *J. Mol. Biol.* **72**, 163 (1972).

270. A. W.-M. Lee and M. Karplus, *Proc. Natl. Acad. Sci. USA* **80**, 7055 (1983).

271. B. R. Gelin and M. Karplus, *Proc. Natl. Acad. Sci. USA* **74**, 801 (1977).

272. L. Anderson, *J. Mol. Biol.* **79**, 495 (1973).

273. A. W.-M. Lee and M. Karplus (to be published).

274. E. Henry, M. Levitt, and W. A. Eaton, *Proc. Natl. Acad. Sci. USA* **82**, 2034 (1985).

275. J. L. Martin, A. Migus, C. Poyart, Y. Lecarpentier, R. Astier, and M. Antonetti, *Proc. Natl. Acad. Sci. USA* **80**, 173 (1983).

276. R. Kretsinger, *CRC Crit. Rev. Biochem.* **8**, 119 (1980).

277. D. Herzberg, J. Moult, and M. N. G. James, *J. Biol. Chem.* **261**, 2638 (1986).

278. W. Bode and R. Huber, *Acc. Chem. Res.* **11**, 114 (1978).

279. T. Alber, D. W. Banner, A. C. Bloomer, G. A. Petsko, D. Phillips, P. S. Rivers, and I. A. Wilson, *Phil. Trans. Roy. Soc. London* **B293**, 159 (1981).

280. M. N. G. James and A. Sielecki, *J. Mol. Biol.* **163**, 299 (1983).

281. A. Klug and A. Durham, *Cold Spring Harbor Symp. Quant. Biol.* **36**, 449 (1971).

282. A. Klug, *Harvey Lect.* **74**, 141 (1979).

283. C. O. Pabo and M. Lewis, *Nature (Lond.)* **298**, 443 (1982).

284. M. A. Weiss, R. T. Sauer, D. J. Patel, and M. Karplus, *Biochemistry* **23**, 5090 (1984).

285. D. C. Wiley and J. J. Skehel, *Ann. Rev. Biochem.* **56**, 365 (1987).

286. R. Rigler and W. Wintermeyer, *Ann. Rev. Biophys. Biophys. Chem.* **12**, 475 (1983).

287. C. Boesch, A. Bundi, M. Oppliger, and K. Wüthrich, *Eur. J. Biochem.* **91**, 209 (1978).

288. W. Braun, G. Wider, K. H. Lee, and K. Wüthrich, *J. Mol. Biol.* **169**, 921 (1983).

289. M. Wagman, C. M. Dobson, and M. Karplus, *FEBS Lett.* **119**, 265 (1980).

290. K. Sasaki, S. Dockerill, D. A. Adamiak, I. J. Tickle, and T. Blundell, *Nature (Lond.)* **257**, 751 (1975).

291. M. Levitt, *J. Mol. Biol.* **104**, 59 (1976).

292. S. Inoue, T. Sano, Y. Yakabe, H. Ushio, and T. Yasunaga, *Biopolymers* **18**, 681 (1979).

293. C. B. Anfinsen and H. A. Scheraga, *Adv. Protein Chem.* **29**, 205 (1975).

294. P. S. Kim and R. L. Baldwin, *Ann. Rev. Biochem.* **51**, 459 (1982).

295. J. E. Brown and W. A. Klee, *Biochemistry* **10**, 470 (1971).

296. R. M. Epand, *J. Biol. Chem.* **247**, 2132 (1972).

297. G. G. Hammes and S. E. Schullery, *Biochemistry* **7**, 3882 (1968).

298. W. B. Gratzer and G. H. Beaven, *J. Biol. Chem.* **244**, 6675 (1969).

299. B. Panijpan and W. B. Gratzer, *Eur. J. Biochem.* **45**, 547 (1974).

300. K. R. Shoemaker, P. S. Kim, E. J. York, J. M. Stewart, and R. L. Baldwin, *Nature* **326**, 563 (1987).

301. P. K. Ponnuswamy, P. K. Warme, and H. A. Scheraga, *Proc. Natl. Acad. Sci. USA* **70**, 830 (1973).

302. M. Karplus and D. Weaver, *Nature (Lond.)* **260**, 404 (1976).

303. M. Karplus and D. Weaver, *Biopolymers* **18**, 1421 (1979).

304. D. L. Weaver, *Biopolymers* **23**, 675 (1984).

305. C. Ghelis, M. Tempete-Gaillourdet, and J. Yon, *Biochem. Biophys. Res. Commun.* **84**, 31 (1978).

306. D. Bashford, D. L. Weaver, and M. Karplus, *J. Biomol. Struct. Dyn.* **1**, 1243 (1984).

307. S. Y. Lee and M. Karplus, *J. Phys. Chem.* **92**, 1075 (1988).

308. M. G. Munowitz, C. M. Dobson, R. G. Griffin, and S. C. Harrison, *J. Mol. Biol.* **141**, 327 (1980).

309. T. Butz, A. Lerf, and R. Huber, *Phys. Rev. Lett.* **48**, 890 (1982).

310. P. B. Sigler, D. M. Blow, B. W. Matthews, and R. Henderson, *J. Mol. Biol.* **35**, 143 (1968).

311. A. Brünger, M. Karplus, and R. Huber, *Biochemistry* **26**, 5153 (1987).

312. C. F. Wong and J. A. McCammon, *Isr. J. Chem.* **27**, 211 (1987).

313. D. W. Banner, A. C. Bloomer, G. A. Petsko, D. C. Phillips, C. I. Pogson, and I. A. Wilson, *Nature (Lond.)* **255**, 609 (1975).

314. T. Alber, W. A. Gilbert, D. Ringe, and G. A. Petsko, in Ref. 41.

314a. D. Joseph, G. A. Petsko, and M. Karplus (in progress).

315. J. Stackhouse, K. P. Nambiar, J. J. Burbaum, D. M. Stauffer, and S. A. Benner, *J. Am. Chem. Soc.* **107**, 2757 (1985).

316. W. Kauzmann, *Adv. Protein Chem.* **14**, 1 (1959).

317. C. L. Brooks III and M. Karplus, *Methods Enzymol.* **127**, 369 (1986).

318. D. Beece, L. Eisenstein, H. Frauenfelder, D. Good, M. C. Marden, L. Reinisch, A. H. Reynolds, L. B. Sorensen, and K. T. Yue, *Biochemistry* **19**, 5147 (1980).

319. J. L. Finney, in *Water: A Comprehensive Treatise,* Vol. 6, F. Franks, ed., Plenum Press, New York, 1979, p. 47.

320. S. J. Wodak, P. Alard, P. Delhaise, and C. Renneboog-Squilbin, *J. Mol. Biol.* **181**, 317 (1985).

321. J. Hermans, H. J. C. Berendsen, W. F. van Gunsteren, and J. P. M. Postma, *Biopolymers* **23**(8), 1513 (1984).

322. H. J. C. Berendsen, W. F. van Gunsteren, H. R. S. Zwindermann, and R. G. Geurten, *Ann. New York Acad. Sci.* **482**, 269 (1986).

323. H-ai Yu and M. Karplus (unpublished).

324. T. Ichiye and M. Karplus, *Biochemistry* **22**, 2884 (1983).

325. M. Rholam, S. Scarlata, and G. Weber, *Biochemistry* **23**, 6793 (1984).

326. O. B. Ptitsyn, *FEBS Lett.* **93**, 1 (1978).

327. B. Gavish, in *The Fluctuating Enzyme,* G. R. Welsh, ed., Wiley-Interscience, New York, 1986, p. 263.

328. M. Karplus, *Adv. Biophys.* **18**, 165 (1984).

329. R. M. Levy, M. Karplus, and J. A. McCammon, *J. Am. Chem. Soc.* **103**, 994 (1981).

330. B. M. Pettitt and M. Karplus (to be published).

331. F. M. Richards, *Ann. Rev. Biophys. Bioeng.* **6**, 151 (1977).

332. R. F. Grote and J. T. Hynes, *J. Chem. Phys.* **74**, 4465 (1981).

333. G. Lipari, A. Szabo, *J. Am. Chem. Soc.* **104**, 4546 (1982).

334. S. Y. Lee and M. Karplus, *J. Chem. Phys.* **81**, 6106 (1984).

335. E. Helfaud, Z. R. Wasserman, and T. A. Weber, *Macromolecules,* **13**, 526 (1980).

336. J. A. Hartsuck and W. N. Lipscomb, *The Enzymes* **3**, 3rd ed. 1 (1971).

337. S. J. Gardell, C. S. Craik, D. Hilvert, M. S. Urdea, and W. J. Rutter, *Nature* **317**, 551 (1985).

338. R. M. Levy, M. Karplus, and J. A. McCammon, *Chem. Phys. Lett.* **65**, 4 (1979).

339. L. R. Pratt and D. Chandler, *J. Chem. Phys.* **67**, 3683 (1977).

340. J. A. McCammon, O. A. Karim, T. P. Lybrand, and C. F. Wong, *Ann. N.Y. Acad. Sci.* **482**, 210 (1987).

341. H. J. C. Berendsen, J. P. M. Postma, W. F. van Gunsteren, and J. Hermans, in *Intermolecular Forces,* B. Pullman, ed., Reidel, Dordrecht, 1981.

342. M. Mezei, P. K. Mehrotra, and D. L. Beveridge, *J. Am. Chem. Soc.* **107**, 2239 (1985).

343. W. L. Jorgensen, *J. Chem. Phys.* **77**, 4156 (1982).

344. C. Pangali, M. Rao, and B. J. Berne, *J. Chem. Phys.* **71**, 2975 (1979).

345. A. Ben-Naim, *Water, A Comprehensive Treatise: Introduction to a Molecular Theory,* Vol. II Plenum Press, New York, 1973.

346. J. Brady and M. Karplus (to be published).

347. J. Deisenhofer and W. Steigemann, *Acta Cryst.* **B31**, 238 (1975).

348. A. A. Kossiakoff, *Nature (Lond.)* **296**, 713 (1982).

349. C. C. F. Blake, W. C. A. Pulford, and P. J. Artymiuk, *J. Mol. Biol.* **167**, 693 (1983).

350. (a) W. A. Gilbert, A. L. Fink, and G. A. Petsko, *Biochemistry* (in press); (b) R. L. Campbell and G. A. Petsko, *Biochemistry* (in press).

351. F. M. Richards and H. W. Wyckhoff, in *The Enzymes,* 3rd ed., Vol. 4, Academic Press, New York, 1971, p. 647.

352. P. Blackburn and S. Moore, in *The Enzymes,* 3rd ed., Vol. 15, Academic Press, New York, 1982, p. 317.

353. G. C. K. Roberts, E. A. Dennis, D. H. Meadows, J. S. Cohen, and O. Jardetzky, *Proc. Natl. Acad. Sci. USA* **62,** 1151 (1969).

354. J. B. Matthew and F. M. Richards, *Biochemistry* **21,** 4989 (1982).

355. (a) P. J. Desmeules and L. C. Allen, *J. Chem. Phys.* **72,** 4731 (1980); (b) W. Reiher and M. Karplus (unpublished results).

356. P. Kebarle, *Ann. Rev. Phys. Chem.* **28,** 445 (1977).

357. A. Wlodawer, in *Biological Macromolecules and Assemblies:* Vol. 2, Nucleic Acids and Interactive Proteins, Wiley, New York, 1985, pp. 394–439.

358. E. N. Baker and R. E. Hubbard, *Prog. Biophys. Mol. Biol.* **44,** 97 (1984).

359. L. Sawyer and M. N. G. James, *Nature (Lond.)* **295,** 79 (1982).

360. I. Tabushi, Y. Kiyosuke, and K. Yamamura, *J. Am. Chem. Soc.* **103,** 5255 (1981).

361. P. S. Marfey, M. Uziel, and J. Little, *J. Biol. Chem.* **240,** 3270 (1965).

362. P. C. Weber, F. R. Salemme, S. H. Lin, Y. Konishi, and H. A. Scheraga, *J. Mol. Biol.* **181,** 453 (1985).

363. A. Cudd and I. Fridovich, *J. Biol. Chem.* **257,** 11443 (1982).

363a. I. Klapper, R. Hagstrom, R. Fine, K. Sharp, and B. Honig, *Proteins* **1,** 47 (1986).

364. C. L. Brooks III, A. T. Brünger, M. Franc, K. Haydock, L. C. Allen, and M. Karplus, in *International Symposium on Bioorganic Chemistry,* R. Breslow, ed., New York Academy of Sciences, New York, 1986, p. 295.

365. P. Debye, *Trans. Electrochem. Soc.* **82,** 265 (1942).

366. G. Lamm and K. Shulten, *J. Chem. Phys.* **78,** 2713 (1983).

367. O. G. Berg and M. Ehrenberg, *Biophys. Chem.* **17,** 13 (1983).

368. D. Shoup and A. Szabo, *Biophys. J.* **40,** 33 (1982).

369. (a) K. C. Chou and G. P. Zhou, *J. Am. Chem. Soc.* **104,** 1409 (1982); (b) S. A. Allison, G. Ganti, and J. A. McCammon, *Biopolymers* **24,** 1323 (1985).

370. S. H. Northrup, S. A. Allison, and J. A. McCammon, *J. Chem. Phys.* **80,** 1517 (1984).

371. S. A. Allison, N. Srinivasan, J. A. McCammon, and S. H. Northrup, *J. Phys. Chem.* **88,** 6152 (1984).

371a. E. Dickinson, *Chem. Soc. Rev.* **14,** 421 (1985).

372. M. Salin and W. Wilson, *Mol. Cell. Biochem.* **36,** 157 (1981).

373. J. W. van Leeuwen, *FEBS Lett.* **156,** 262 (1983).

374. E. Getzoff, J. Tainer, P. Weiner, P. Kollman, J. Richardson, and D. Richardson, *Nature (Lond.)* **306,** 287 (1983).

375. A. Beyer, C. L. Brooks, and M. Karplus (to be published).

376. T. Head-Gordon and C. L. Brooks III, *J. Phys. Chem.* **91,** 3342 (1987).

377. H. Frauenfelder, G. A. Petsko, and D. Tsernoglu, *Nature (Lond.)* **380,** 558 (1978).

378. C. Post and M. Karplus, *J. Am. Chem. Soc.* **108,** 1317 (1986).

379. C. F. Wong and J. A. McCammon, *J. Am. Chem. Soc.* **108,** 3830 (1986).

380. J. Brady and M. Karplus, *J. Am. Chem. Soc.* **107**, 6103 (1985).

380a. B. M. Pettitt, P. R. Rossky, and M. Karplus, *J. Phys. Chem.* **90**, 6335 (1986).

381. A. G. Szent-Györgyi and C. Cohen, *Science* **126**, 697 (1957).

382. D. R. Davies, *J. Mol. Biol.* **9**, 605 (1964).

383. P. Y. Chou and G. D. Fasman, *Adv. Enzymol.* **45**, 45 (1978), and references therein.

384. S. Tanaka and H. A. Scheraga, *Macromolecules* **9**, 168 (1976).

385. F. Parak and E. W. Knapp, *Proc. Natl. Acad. Sci. USA* **81**, 7088 (1984).

386. J. A. Rupley, E. Gratton, and G. Careri, *TIBS* **8**, 18 (1983).

387. P. L. Privalov, *Adv. Protein Chem.* **33**, 167 (1979).

388. J. O. Hutchens, A. G. Cole, and J. W. Stout, *J. Biol. Chem.* **244**, 26 (1969).

389. M. Karplus, T. Ichiye, and B. M. Pettitt, *Biophys. J.* **52**, 1083 (1987).

390. G. P. Singh, H. J. Schink, H. V. Loehneysen, F. Parak, and S. Hunklinger, *Z. Phys.* **B55**, 23 (1984).

391. B. Jacrot, S. Cusack, A. J. Dianoux, and D. M. Engelman, *Nature (Lond.)* **300**, 84 (1982).

392. R. Hawkes, M. G. Grutter, and J. Schellman, *J. Mol. Biol.* **175**, 195 (1984).

392a. T. Alber, S. Dao-pin, K. Wilson, J. A. Wozuik, S. P. Cook, and B. M. Matthews, *Nature* **330**, 41 (1987).

393. D. Shortle and A. K. Meeker, *Proteins* **1**, 81 (1986).

394. M. Mezei, S. Swaninathan, and D. L. Beveridge, *J. Am. Chem. Soc.* **100**, 3255 (1978).

395. B. Tembe and J. A. McCammon, *Comput. & Chem.* **8**, 281 (1984).

396. T. P. Lybrand, J. A. McCammon and G. Wipff, *Proc. Natl. Acad. Sci. USA* **83**, 833 (1986).

397. J. A. McCammon, T. P. Lybrand, S. A. Allison, and S. H. Northrup, in *Biomolecular Stereodynamics* III, R. H. Sarma, ed., Adenine Press, New York, 1986, pp. 227–236.

398. K. Zakrzewska, R. Lavery, and B. Pullman, *Nucleic Acid Res.* **12**, 6559 (1984).

398a. P. A. Bash, U. C. Singh, R. Langridge, and P. A. Kollman, *Science,* **236**, 564 (1987).

399. J. Chandrasekhar, S. F. Smith, and W. L. Jorgensen, *J. Am. Chem. Soc.* **106**, 3049 (1984).

400. (a) W. Albery and M. Kreevoy, *Adv. Phys. Org. Chem.* **16**, 87 (1978); (b) S. S. Shaik, *J. Am. Chem. Soc.* **106**, 1227 (1984).

401. A. Warshel and F. Sussman, *Proc. Natl. Acad. Sci. USA* **83**, 3806 (1986).

402. S. J. Weiner, G. L. Seibel, and P. A. Kollman, *Proc. Natl. Acad. Sci.* **83**, 649 (1986).

403. T. Takano, *J. Mol. Biol.* **110**, 537 (1977).

404. B. P. Schoenborn, H. C. Watson, and J. C. Kendrew, *Nature (Lond.)* **207**, 18 (1965).

405. R. F. Tilton, I. D. Kuntz, and G. A. Petsko, *Biochemistry* **23**, 2849 (1984).

406. I. D. Kuntz and R. F. Tilton, *Biochemistry* **21**, 6850 (1982).

407. J. Hermans and S. Shankar, *Isr. J. Chem.* **27**, 225 (1986).

407a. B. Tidor and M. Karplus (to be published).

408. W. van Gunsteren, private communication.

409. W. L. Jorgensen and C. Ravimohan, *J. Chem. Phys.* **83**, 3050 (1985).

410. G. Wipff, A. Dearing, P. K. Weiner, J. M. Blaney, and P. A. Kollman, *J. Am. Chem. Soc.* **105**, 997 (1983).

410a. A. Carotti, C. Hansch, M. M. Mueller, and J. M. Blaney, *J. Med. Chem.* **27**, 1401 (1984).

410b. S. P. Gupta, *Chem. Rev.* **87**, 1183 (1987).

410c. See the series *Topics in Molecular Pharmacology* (e.g., ref. 26a).

411. (a) H. P. Weber, T. Lybrand, U. Singh, and P. A. Kollman, *J. Mol. Graph.* **4**, 56 (1986); (b) T. Lybrand, Ph.D. thesis, University of California, San Francisco, 1984.

412. (a) L. Kao, V. J. Hruby, B. M. Pettitt, and M. Karplus, *J. Am. Chem. Soc.* **110**, 3351 (1988); (b) B. M. Pettitt, V. J. Hruby, and M. Karplus (to be published).

413. U. H. Zucker and H. Schulz, *Acta Cryst.* **A38**, 563 (1982).

414. J. Kuriyan, M. Karplus, and G. A. Petsko, *Proteins* **2**, 1 (1987).

415. S. E. V. Phillips, *J. Mol. Biol.* **142**, 531 (1980).

416. J. H. Konnert and W. A. Hendrickson, *Acta Cryst.* **A36**, 344 (1980).

417. I. Glover, I. Haneef, J. Pitts, S. Wood, D. Moss, I. Tickle, and T. Blundell, *Biopolymers* **22**, 293 (1983).

418. P. J. Artymiuk, C. C. F. Blake, D. E. P. Grace, S. J. Oatley, D. C. Phillips, and M. J. E. Sternberg, *Nature (Lond.)* **280**, 563 (1979).

419. H. Yu, M. Karplus, and W. A. Hendrickson, *Acta Cryst.* **B41**, 191 (1985).

420. T. Blundell et al., in Ref. 42, p. 85.

420a. A. T. Brünger, J. Kuriyan, and M. Karplus, *Science* **235**, 459 (1987).

420b. S. Kirkpatrick, C. D. Gelatt, Jr., and M. P. Vecchi, *Science* **220**, 671 (1983).

421. J. H. Konnert and W. A. Hendrickson, *Acta Cryst.* **36**, 344 (1980).

422. R. E. London, in *Magnetic Resonance in Biology*, Vol. 1, J. S. Cohen, ed., Wiley, New York, 1980, p. 1.

423. R. M. Levy, M. Karplus, and P. G. Wolynes, *J. Am. Chem. Soc.* **103**, 5998 (1981).

424. I. D. Campbell and C. M. Dobson, *Methods Biochem. Anal.* **25**, 1 (1979).

425. E. T. Olejniczak, C. M. Dobson, M. Karplus, and R. M. Levy, *J. Am. Chem. Soc.* **106**, 1923 (1984).

426. I. Solomon, *Phys. Rev.* **99**, 559 (1955).

427. R. M. Levy, C. M. Dobson, and M. Karplus, *Biophys. J.* **39**, 107 (1982).

428. G. Lipari and A. Szabo, *J. Am. Chem. Soc.* **104**, 4559 (1982).

429. K. Wüthrich, G. Wagner, R. Richarz, and W. Braun, *Biophys. J.* **32**, 549 (1980).

430. G. Lipari, A. Szabo, and R. M. Levy, *Nature (Lond.)* **300**, 197 (1982).

431. E. T. Olejniczak, F. M. Poulsen, and C. M. Dobson, *J. Am. Chem. Soc.* **103**, 6574 (1981).

432. R. J. Wittebort and A. Szabo, *J. Chem. Phys.* **69**, 1722 (1978).

433. R. E. London and J. Avitabile, *J. Am. Chem. Soc.* **100**, 7159 (1978).

434. F. M. Poulsen, J. C. Hoch, and C. M. Dobson, *Biochemistry* **19**, 2597 (1980).

435. T. F. Havel and K. Wüthrich, *J. Mol. Biol.* **182**, 281 (1985).

436. R. Kaptein, E. R. P. Zuiderweg, R. M. Scheek, R. Boelens, and W. F. van Gunsteren, *J. Mol. Biol.* **182**, 179 (1985).

437. G. M. Clore, A. M. Gronenborn, A. T. Brünger, and M. Karplus, *J. Mol. Biol.* **186**, 435 (1985).

438. A. T. Brünger, G. M. Clore, A. M. Gronenborn, and M. Karplus, *Proc. Natl. Acad. Sci. USA* **83**, 3801 (1986).

439. A. T. Brünger, R. L. Campbell, G. M. Clore, A. M. Gronenborn, M. Karplus, G. A. Petsko, and M. M. Teeter, *Science* **235**, 458 (1987).

440. I. D. Campbell, C. M. Dobson, G. R. Moore, S. J. Perkins, and R. J. P. Williams, *FEBS Lett.* **70**, 96 (1976).

441. W. van Gunsteren and M. Karplus, *Nature (Lond.)* **293**, 677 (1981).

442. I. Ghosh, J. A. McCammon, *Biophys. J.* **51**, 637 (1987).

443. L. Blumenfeld, *J. Theoret. Biol.* **58**, 269 (1976).

444. G. Wagner, *FEBS Lett.* **112**, 280 (1980).

445. A. Zipp and W. Kauzmann, *Biochemistry* **12**, 4217 (1973).

446. M. Karplus and J. A. McCammon, *FEBS Lett.* **121**, 34 (1981).

447. C. E. Johnson and F. A. Bovey, *J. Chem. Phys.* **29**, 1012 (1958).

448. B. Pullman and C. Giessner-Prettre, *J. Theor. Biol.* **31**, 287 (1971).

449. M. Karplus, *J. Am. Chem. Soc.* **85**, 2870 (1963).

450. S. J. Perkins, *Biol. Magn. Reson.* **4**, 193 (1982).

451. J. C. Hoch, C. M. Dobson, and M. Karplus, *Biochemistry* **21**, 1118 (1982).

452. J. C. Hoch, C. M. Dobson, and M. Karplus, *Biochemistry* **24**, 3831 (1985).

453. D. Wallach, *J. Chem. Phys.* **47**, 5258 (1967).

454. T. Tao, *Biopolymers* **8**, 609 (1969).

454a. A. Szabo, *J. Chem. Phys.* **81**, 150 (1984).

455. J. R. Lakowicz, B. P. Maliwal, H. Cherek, and A. Balter, *Biochemistry* **22**, 1741 (1983).

456. I. Munro, I. Pecht, and L. Stryer, *Proc. Natl. Acad. Sci. USA* **76**, 56 (1979).

457. R. M. Levy and A. Szabo, *J. Am. Chem. Soc.* **104**, 2073 (1982).

458. A. Kasprzak and G. Weber, *Biochemistry* **21**, 5924 (1982).

459. J. R. Lakowicz and G. Weber, *Biochemistry* **12**, 4161 (1973).

459a. L. X. Q. Chen, R. A. Eugh, A. T. Brünger, D. T. Nguyen, M. Karplus, and G. R. Fleming, *Biochemistry* (in press).

460. T. Rizzo, Y. D. Park, L. Peteanu, and D. Levy, *J. Chem. Phys.* **83**, 4819 (1985).

461. J. L. Martin, private communication.

462. B. Hudson, private communication.

463. K. Itoh and T. Shimanouchi, *Biopolymers* **9**, 383 (1970).

464. W. L. Peticolas, *Biopolymers* **18**, 747 (1979).

465. A. Dwivedi and S. Krimm, *Biopolymers* **23**, 923 (1984).

466. A. Dwivedi and S. Krimm, *Macromolecules* **15**, 186 (1982).

467. N.-T. Yu, *CRC Crit. Rev. Biochem.* **4**, 229 (1977).

468. T. G. Spiro and B. P. Gaber, *Ann. Rev. Biochem.* **46**, 553 (1977).

469. H. Brunner and H. Sussner, *Biochim. Biophys. Acta* **271**, 16 (1972).

470. K. A. Bagley, V. Balogh-Nair, A. A. Croteau, G. Dollinger, T. G. Ebrey, L. Eisenstein, M. K. Hong, K. Nakanishi, and J. Vittitow, *Biochemistry* **24**, 6055 (1985).

471. T. G. Spiro, in *Iron Porphyrins,* A. B. P. Lever and H. B. Gray, eds., Addison-Wesley, Reading, Mass., 1983.

472. M. R. Ondrias, J. M. Friedman, and D. L. Rousseau, *Science* **220**, 615 (1983).

473. L. Genzel, F. Keilmann, T. P. Martin, G. Winterling, Y. Yacoby, H. Fröhlich, and M. W. Makinen, *Biopolymers* **15**, 213 (1976).

474. M. Ataka and S. Tanaka, *Biopolymers* **18**, 507 (1979).

475. W. Drexel and W. L. Peticolas, *Biopolymers* **14**, 715 (1975).

476. H. D. Middendorf, *Ann. Rev. Biophys. Bioeng.* **13**, 425 (1984).

477. B. K. Alfred, G. C. Stirling, J. W. White, *Faraday Symp. Chem. Soc.* **6**, 135 (1972).

478. H. Hervet, A. J. Dianoux, R. E. Lechner, and F. Volino, *J. Phys. (Paris)* **37**, 587 (1976).

479. H. D. Bartunik, P. Jolles, J. Berthou, and A. J. Dianoux, *Biopolymers* **21**, 43 (1982).

480. H. D. Middendorf, in *Neutrons in Biology*, B. P. Schoenborn, ed., Plenum Press, New York, 1984.

481. A. C. Zemach and R. J. Glauber, *Phys. Rev.* **101**, 118 (1956).

482. S. W. Lovesey, *Theory of Neutron Scattering from Condensed Matter*, Clarendon Press, New York, 1984.

483. J. Smith, S. Cusack, U. Pezzeca, B. Brooks, and M. Karplus, *J. Chem. Phys.* **85**, 3636 (1986).

484. S. Cusack, J. Smith, and J. L. Finney (to be published).

485. S. Cusack, J. Smith, J. L. Finney, B. Tidor, and M. Karplus, *J. Mol. Biol.* (in press).

486. A. Abragam and B. Bleaney, *Electron Paramagnetic Resonance of Transition Ions*, Oxford University Press, New York, 1970.

487. I. Bertini, F. Briganti, S. H. Koenig, and C. Luchinat, *Biochemistry* **24**, 6287 (1985).

488. R. C. Herrick and H. J. Stapelton, *J. Chem. Phys.* **65**, 4778 (1976).

489. G. C. Wagner, J. T. Colvin, J. P. Allen, and H. J. Stapelton, *J. Am. Chem. Soc.* **107**, 5589 (1985).

490. K. W. H. Stevens, *Rep. Prog. Phys.* **30**, 189 (1967).

491. T. Bray, G. C. Brown, Jr., and A. Kiel, *Phys. Rev.* **127**, 730 (1962).

492. S. Alexander and R. Orbach, *J. Phys. (Paris) Lett.* **43**, 625 (1982).

493. R. Elber and M. Karplus, *Phys. Rev. Lett.* **56**, 394 (1986).

494. S. Alexander, J. Bernasconi, W. R. Schneider, and R. Orbach, *Rev. Mod. Phys.* **53**, 175 (1981).

495. L. Finegold and J. L. Cude, *Nature* **238**, 38 (1972).

496. K. Linderstrøm-Lang, *Chem. Soc. (London) Spec. Publ.* **2**, 1 (1955).

497. M. Delepierre, C. M. Dobson, M. A. Howarth, and F. M. Poulsen, *Eur. J. Biochem.* **145**, 389 (1984).

498. G. Wagner and K. Wüthrich, *J. Mol. Biol.* **134**, 75 (1979).

499. B. D. Hilton and C. K. Woodward, *Biochemistry* **17**, 3325 (1978).

500. S. A. Mason, G. A. Bentley, and G. J. McIntyre, *Basic Life Sci.* **27**, 323 (1984).

501. A. Wlodawer and L. Sjölin, *Proc. Natl. Acad. Sci. USA* **79**, 1418 (1982).

502. G. A. Bentley, M. Delepierre, C. M. Dobson, S. A. Mason, F. M. Poulsen, and R. E. Wedin, *J. Mol. Biol.* **170**, 243 (1983).

503. J. B. Matthew and F. M. Richards, *J. Biol. Chem.* **258**, 3039 (1983).

504. E. Tüchsen and C. Woodward, *J. Mol. Biol.* **185**, 421 (1985).

505. M. Delepierre, C. M. Dobson, M. Karplus, F. M. Poulson, D. J. States, and R. E. Wedin, *J. Mol Biol.* **197**, 111 (1987).

505a. M. Levitt, *Nature* **294**, 379 (1981).

505b. G. Wagner and K. Wüthrich, *J. Mol. Biol.* **160**, 343 (1982).

506. R. P. Sheridan, R. M. Levy, and S. W. Englander, *Proc. Natl. Acad. Sci. USA* **80**, 5569 (1983).

507. V. I. Goldanskii and R. H. Herber, *Chemical Applications of Mössbauer Spectroscopy*, Academic Press, New York, 1968.

508. Y. F. Krupyanskii, F. Parak, V. I. Godanskii, R. L. Mössbauer, E. E. Gaubmann, H. Engelmann, and I. P. Suzdalev, Z. Naturforsch. **37C**, 57 (1982).

509. F. Parak, E. W. Knapp, and Kucheida, J. Mol. Biol. **161**, 177 (1982).

510. E. R. Bauminger, S. G. Cohen, I. Nowik, S. Ofer, and J. Yariv, Proc. Natl. Acad. Sci. USA **80**, 736 (1983).

511. F. Parak, E. N. Frolov, A. A. Kononenko, R. L. Mössbauer, V. I. Goldanskii, and A. B. Rubin, FEBS Lett. **117**, 368 (1980).

512. S. G. Cohen, E. R. Bauminger, I. Nowik, S. Ofer, and J. Yariv, Phys. Rev. Lett. **46**, 1244 (1981).

513. G. P. Singh, F. Parak, S. Hunklinger, and S. Dransfeld, Phys. Rev. Lett. **47**, 685 (1981).

514. E. W. Knapp, S. F. Fischer, and F. Parak, J. Chem. Phys. **78**, 4701 (1983).

515. W. Nadler and K. Schulten, Proc. Natl. Acad. Sci. USA **81**, 5719 (1984).

516. W. Nadler, A. T. Brünger, K. Schulten, and M. Karplus, Proc. Nat. Acad. Sci. (USA) **84**, 7933 (1987).

517. L. J. Parkhurst, Ann. Rev. Phys. Chem. **30**, 503 (1979).

518. H. Frauenfelder, in Structure, Dynamics and Function of Biomolecules, L. Nilsson, ed., Springer Series in Biophysics, **I**, 10 (1987).

519. N. Agmon and J. J. Hopfield, J. Chem. Phys. **79**, 2042 (1983).

520. W. Bialek and R. F. Goldstein, Biophys. J. **48**, 1027 (1985).

521. E. R. Henry, J. H. Sommer, J. Hofrichter, and W. A. Eaton, J. Mol. Biol. **166**, 443 (1983).

522. J. M. Friedman, D. L. Rousseau, and M. R. Ondrias, Ann. Rev. Phys. Chem. **33**, 471 (1982).

523. D. L. Rousseau and P. V. Argade, Proc. Natl. Acad. Sci. USA **83**, 1310 (1986).

524. A. Ansari, J. Berendzen, S. F. Bowne, H. Frauenfelder, I. E. T. Iben, T. B. Sauke, E. Shyamsunder, and R. D. Young, Proc. Natl. Acad. Sci. USA **82**, 5000 (1985).

525. H. Frauenfelder, in Structure and Motion, E. Clementi, G. Corongiu, M. H. Sarma, and R. H. Sarma, eds., Adenine Press, Guilderland, N.Y., 1985, p. 205.

526. T. L. Hill, Thermodynamics of Small Systems, Part 1, W. A. Benjamin, Inc., New York, 1963.

527. D. H. J. Mackay, P. H. Berens, K. R. Wilson, and A. T. Hagler, Biophys. J. **46**, 229 (1983).

528. B. Roux and M. Karplus, Biophys. J., **53**, 297 (1988).

529. N. Liang, G. J. Pielak, A. G. Mauk, M. Smith, and B. M. Hoffman, Proc. Natl. Acad. Sci. **84**, 1249 (1987).

530. J. Deisenhofer, O. Epp, K. Miki, R. Huber, and H. Michel, J. Mol. Biol. **180**, 385 (1984).

530a. J. P. Allen, G. Feher, T. O. Yeates, D. C. Rees, J. Deisenhofer, H. Michel, and R. Huber, Proc. Nat. Acad. Sci. USA **83**, 8589 (1986).

531. R. A. Marcus and N. Sutin, Biochim. Biophys. Acta **811**, 265 (1985).

532. M. Bixon and J. Jortner, J. Phys. Chem. **90**, 3795 (1986).

533. G. R. Fleming, J. L. Martin, J. Breton, Nature **333**, 192 (1988).

INDEX

251